Inorganic Chemistry Concepts
Volume 5

Editors

Christian K. Jørgensen, Geneva · Michael F. Lappert, Brighton
Stephen J. Lippard, New York · John L. Margrave, Houston
Kurt Niedenzu, Lexington · Heinrich Nöth, Munich
Robert W. Parry, Salt Lake City · Hideo Yamatera, Nagoya

Takeshi Tominaga
Enzo Tachikawa

Modern
Hot-Atom Chemistry
and Its Applications

With 57 Figures and 34 Tables

Springer-Verlag
Berlin Heidelberg New York 1981

Prof. Dr. Takeshi Tominaga

Department of Chemistry
Faculty of Science
The University of Tokyo
Hongo, Tokyo/Japan

Dr. Enzo Tachikawa

Japan Atomic Energy
Research Institute
Tokai, Ibaraki/Japan

ISBN-13:978-3-642-68045-8   e-ISBN-13:978-3-642-68043-4
DOI: 10.1007/978-3-642-68043-4

Library of Congress Cataloging in Publication Data.

Tominaga, Takeshi, 1935—. Modern hot-atom chemistry and its applications. (Inorganic chemistry concepts; v. 5.) Includes index. 1. Hot-atom
chemistry. I. Tachikawa, Enzo, 1936—. II. Title. III. Series.
QD601.2.T65   541.3′8   81-5285
ISBN-13:978-3-642-68045-8 (U.S.) AACR2

© Springer-Verlag Berlin, Heidelberg 1981
Softcover reprint of the hardcover 1st edition 1981

2152/3020—543210

Preface

Hot-atom chemistry is a unique field of chemistry dealing with highly excited chemical species resulting from nuclear reactions or radioactive decay processes. Modern hot-atom chemistry includes a broad range of disciplines such as fundamental studies from physical chemistry of gas-phase energetic reactions to inorganic solid-state chemistry, as well as recent practical applications in life sciences and energy-related research.

In spite of the importance of hot-atom chemistry and its applications, its relevance to the other fields of chemistry and related disciplines has attracted little attention and only books and review articles for dedicated hot-atom chemists have been published to date. In this volume, we illustrate the essential aspects of modern hot-atom chemistry for *non-specialists*, with considerable emphasis on its applications in the related fields. We sincerely hope that this volume can promote mutual understanding and collaboration between hot-atom chemists and researchers in other disciplines.

After a brief introduction (Chap. 1) the 2nd chapter gives the non-specialist an idea of experimental techniques commonly used for the production and analysis of hot chemical species.

In Chap. 3, we have explained the concepts of hot-atom reactions in gas, liquid and solid phases with typical examples rather than a comprehensive review of the literature. In view of the current state of accomplishment, the greater part of this chapter is concerned with gas phase studies. Regarding the solid-phase hot-atom chemistry, we have confined ourselves only to introducing *new* concepts and discussing *modern* aspects.

The last chapter is devoted to various applications of hot-atom chemistry in inorganic, analytical and geochemistry and other fields, together with selected subjects of interest possibly related to the future scope of hot-atom chemistry, such as NEET, laser isotope separation and mesic chemistry. We wish to emphasize the significance of the applications because they will certainly promote interdisciplinary collaboration and expand the scope for future developments of hot-atom chemistry.

Acknowledgements are extended to the authors and publishers who generously permitted us to include their figures and tables in this book.

Tokyo and Tokai, May 1981   Takeshi Tominaga
                                                         Enzo Tachikawa

Contents

1 Indroduction

When an atom has undergone nuclear transformation (nuclear reaction or radio-active decay), it often acquires high kinetic energy, or high electric charge. The atom formed with an energy well in excess of the ambient thermal energy (not necessarily via a nuclear event), or highly charged is called a *hot atom*. Such atoms then dissipate most of their kinetic energy in the medium producing radiolytic changes in the surrounding system (those changes will be studied in radiation chemistry) and eventually become stabilized through chemical reactions. If the hot atom is radioactive, we can follow its fate until it becomes chemically combined. Thus *hot-atom chemistry* is a field of studies regarding the fate of the highly excited atoms and molecules mainly resulting from nuclear transformations.

Although physics of the energetic recoiling atoms had been known, the study of hot-atom chemistry was initiated nearly fifty years ago by Szilard and Chalmers [1.1]. They found that when ethyl iodide was irradiated with thermal neutrons some of the radioactive iodine atoms produced by the $^{127}I(n, \gamma)^{128}I$ reaction broke the C-I bond and could be separated into water by simply shaking the irradiated ethyl iodide with water.

While the final goal of hot-atom chemistry is the elucidation of the mechanisms of reactions involving highly excited chemical species, more or less empirical approaches were employed in its early days with appreciable interest in its applications such as the production of enriched radioisotopes and labeled com-pounds. Gas-phase hot reactions were investigated extensively by means of radiogas chromatography, and condensed-phase reactions by other conventional techniques mainly based on wet chemistry.

A renaissance has been brought to classical hot-atom chemistry by the recent progress in experimental techniques such as charge spectrometry and Mössbauer spectroscopy, to name a few. The modern hot-atom chemistry covers wide fields of fundamental research ranging from theoretical approaches and chemical dynamics of high energy reactions in gas-phase to solid-state chemistry of in-organic systems. Since it deals with the reactions of non-Boltzmann systems, gas-phase hot-atom chemistry is complementary to conventional chemical studies of the systems in thermal equilibrium, and greatly contributes to the understan-ding of chemical reactions taking place in the gas phase over a wide energy scale. More recently, hot-atom chemistry has found important applications in the life sciences (nuclear medicine-radiopharmaceuticals and biophysics) and energy-related research (tritium and fission products chemistry in reactors).

In view of the interdisciplinary nature of this field, we have felt it essentially important to direct our attention to applications of hot-atom chemistry in various related fields. Accordingly, its typical applications in inorganic, analytical and geochemistry, physical chemistry, biochemistry, and nuclear medicine, and energy-related research have been discussed in this volume, together with some related subjects (e.g. lasers/NEET and isotope separation, mesic chemistry) which should deserve ample attention of hot-atom chemists in seeking for future scope for developments in this field.

Because of the limited room in this volume, we have tried to explain the reader the present status of hot-atom chemistry with typical examples rather than to provide a complete bibliography. Those who ask for a more detailed record of progress in this field may be advised to refer to relevant books and reviews for specialists [1.2—5], proceedings or status report of international symposia [1.6], or original papers.

References

[1.1] Szilard, L., Chalmers, T. A.: Nature *134*, 462 (1934)
[1.2] Harbottle, G., Maddock. A. G. (eds.): *Chemical Effects of Nuclear Transformations in Inorganic Systems*. North-Holland, 1979
[1.3] Tang, Y. N.: in *Isotopes in Organic Chemistry*. E. Buncel, C. C. Lee (eds.). *4*, 85, Elsevier 1978
[1.4] Urch, D. S.: MTP Int. Rev. Sci., Inorg. Chem. Ser. One, *8*, 149 (1972); ibid. Two, *8*, 49 (1975)
[1.5] Rowland, F. S.: MTP Int. Rev. Sci., Phys. Chem. Ser. One *9*, 109, (1972)
[1.6] *Hot Atom Chemistry Status Report*, International Atomic Energy Agency, Vienna, 1975

2 Experimental Techniques

2.1 Production of Energetic Atoms

Various methods have been used for the study of energetic atoms, e.g. beam methods, photolysis, radiolysis, radioactive decay induced reactions and nuclear recoil. Each method has its own characteristics, yet at the present stage none of them can provide ideal experimental conditions for kinetic research over the entire energy range of interest.

These experimental methods can be classified according to the following order of decreasing experimental difficulty:

1. Crossed, velocity-selected beam experiments, in which only single collisions are allowed.

2. Crossed Maxwellian beams, in which no multiple collisions occur, but the reacting species are not monoenergetic.

3. Controlled initial energy of the reacting species in which multiple collisions of the species are allowed so that the energy state of the reaction is usually not the initial energy.

4. Other cases. Experiments in which the charge state of the particle is unknown or changes, or in which secondary particles can also cause reactions. These types of experiments usually cannot afford quantitative information on reaction kinetics (e.g., decay-induced reaction) although they can permit observation of interesting reactions.

Molecular Beams

The main advantage of this method in chemical research consists in the study of single collision events under well-defined controlled conditions.

The advantages attainable in beam experiments are manifold, although the experimental limitations are also significant, especially in view of the sensitivity of currently available detection systems. As a consequence of this and other technical difficulties, chemical reactions which have been studied successfully by the crossed beam method are very limited.

Siska et al. [2.1.1] have built a crossed beam apparatus, a schematic diagram of which is shown in Fig. 2.1.1, and measured the intermolecular potentials in various reaction systems. Two supersonic nozzle atom beams are allowed to cross at right angles and the atoms scattered in the plane of the beams were detected with a rotatable electron-bombardment mass-filter universal detector.

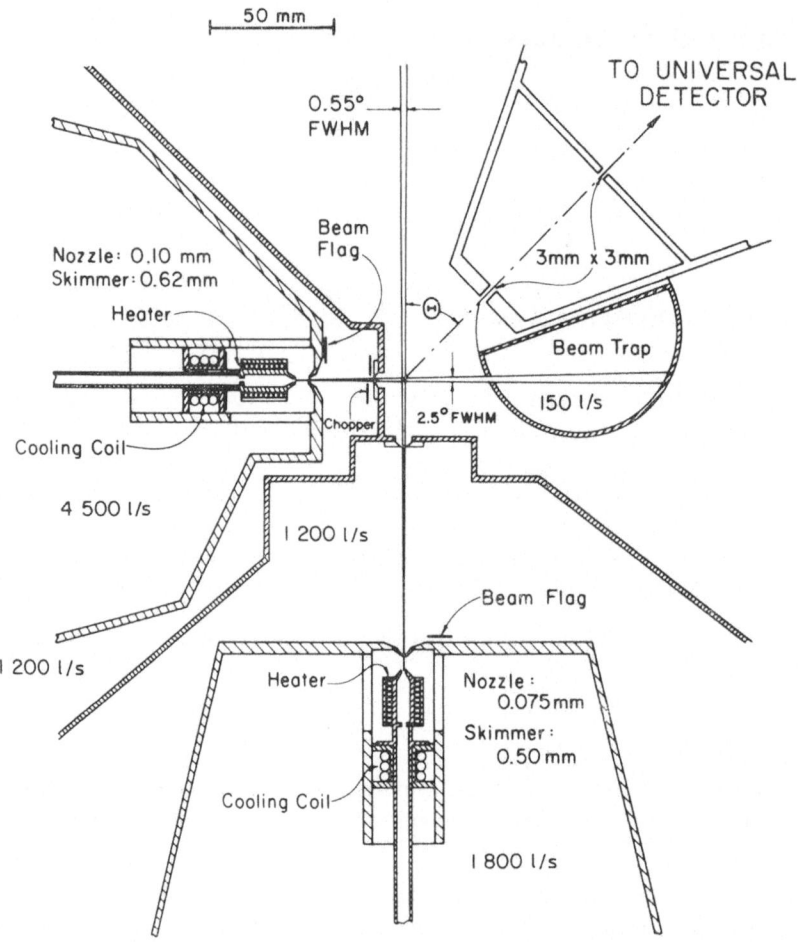

Fig. 2.1.1. Schematic diagram of the crossed-beam arrangement (taken from Ref. 2.1.1)

In the subsequent crossed molecular beam study of the reaction [2.1.2]

$$F + [C_2H_2Cl_2] \rightarrow Cl + [C_2H_2ClF],$$

they found that the recoil energy distribution of the product was compatible with the presence of a reaction complex in which almost all vibrations participate in partitioning the energy.

Herm et al. [2.1.3] have reported the exploratory crossed beam studies regarding the chemistry of gaseous alkaline earth atoms. The measured laboratory distribution of products and derived center-of-mass distribution of recoils are presented for various reaction systems. The results provide important information on the reaction mechanism involved. Fite et al. [2.1.4] have obtained some results on the reactions of D atoms at 2 800 K with H_2 at 77 K to form HD using moderated beams. A crossed beams work has also been published [2.1.5] on T_2 molecules from a chemical accelerator based on the ion acceleration-deceleration-neutralization technique which allows multiple collisions in the collision chamber. In the experiments, T_2 molecules in the 5—100 eV energy range react with n-butane. With 1-butene as a reactant [2.1.6], tritiated 1-butene, n-butane, cis-2- and trans-2-butene are observed. Their yields increase from thresholds below 5 eV to a plateau or broad maximum at high energies, as seen in Fig. 2.1.2.

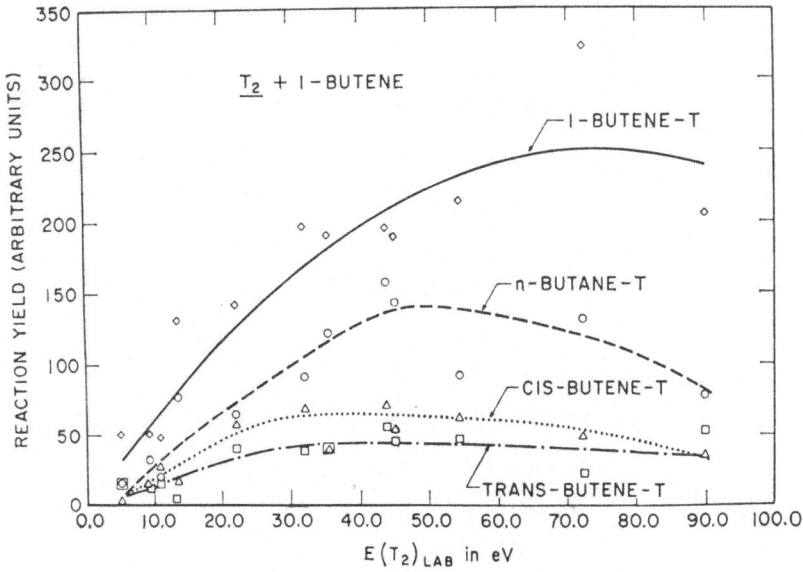

Fig. 2.1.2. Reaction yields of major tritiated products from reactions with 1-butene of T_2 molecules with initial kinetic energies ranging from 5 to 90 eV. (Reprinted with the permission from Ref. 2.1.6. Copyright (1971) American Chemical Society)

Menzinger and Wolfgang [2.1.7] accelerated T^+ and T_2^+ ions to specific energies in the 1—200 eV range. The ions were impinged on a solid c-hexane target. The primary interaction of the ions with the reactant was the neutralization of the ions on the surface of the target; thus, they reacted as neutral tritium atoms. They further extended the work to other alkanes [2.1.8]. Excitation functions $\sigma_i(E)$ for the different reaction channels are derived from the product yields by a kinetic theory analysis. The results indicate that: (1) Hydrogen substitution reaction cross sections are similar for all alkanes studied (n-butane, n-hexane, c-pentane and c-hexane) with a threshold at 1.5 ± 0.5 eV. (2) The reaction cross section for the degraded alkanes shows a strong structure dependence.

Photochemical Means

The photochemical method for the formation of energetic atoms involved the photolysis of a suitable compound, usually hydrogen halides, with ultraviolet light of a selected wavelength. Williams and co-workers [2.1.9] were the first

Table 2.1.1. Production of energetic D atoms by photochemical means (taken from Ref. 2.1.12)

Light source	Parent molecule	Filter	Wavelength absorbed/nm	Initial energy of deuterium atom/(kJ mol^{-1})
Low pressure mercury arc	DBr	^{60}Co irradiated lithium fluoride	184.9	270
Low pressure zinc arc	DBr	cis-2-butene (100 Torr) path length 3 cm	213.8	193
Low pressure cadmium arc	DBr	acrolein in hexane (3×10^{-3} mol l^{-1})	228.8	157
Low pressure mercury arc	DBr	Vycor, effective cutoff at 205 nm	253.7	102
Medium pressure mercury arc	DI	Multilayer-dielectric interference filter—band width at half peak transmittance at 10 nm	313	81
Medium pressure mercury arc	DI	Multilayer-dielectric interference filter—band width at half peak transmittance of 10 nm + 0.6 cm glass u.v. filter	334	57
Thallium arc	DI	Multilayer-dielectric interference filter—band width at half peak transmittance of 10 nm + 0.4 cm Plexiglass u.v. filter	355	40
Medium pressure mercury arc	DI	Multilayer-dielectric interference filter—band width half peak transmittance of 10 nm + 0.4 cm Plexiglass u.v. filter	365	27

to notice the enhanced reactivity of hydrogen atoms formed in photochemical systems. In 1964, Martin and Willard [2.1.10] studied the reaction of 2.9-eV H atoms formed by photochemical dissociation with D_2, CD_4 and C_2D_6.

In the following is considered the UV-absorption of HBr to form an energetic hydrogen atom as an example [2.1.11]. The first absorption of HBr is a broad and smooth continuum in the ultraviolet with a maximum at 178.5 nm. The average initial energy of the resulting hydrogen atoms is calculated from the expression

$$E_L = (m_x/m_{HX}) [E_\lambda - D_0^0(HX) + E_{rot.} + E_{tr.}] \qquad (2.1.1)$$

where m_x and m_{HX} are the masses of a halogen atom and a hydrogen halide molecule; E_λ is the photon energy corresponding to the peak of the source λ distribution and $D_0^0(HX)$ the bond dissociation energy of HX. The average rotational energy for HBr at 300 K is adequately given by the classical approximation, i.e. $E_{rot.} = kT$; also $E_{tr.} = (3/2) kT$. Thus, the kinetic energy of tritium atoms in photodissociation with 1849 A is calculated as 2.8 eV. Similarly,

Fink et al. [2.1.12] produced energetic D atoms by irradiationg DBr or DI with the ultraviolet light of various wavelengths, using suitable filtering systems shown in Table 2.1.1.

However, the energy range covered by this method is limited and cannot be extended towards much higher energies. Another restriction of the method is that any atom heavier than hydrogen cannot be excited suitably to produce energetic atoms.

Nuclear Recoil

Nuclear reactions are accompanied with liberation of a large amount of energy, which can be shared by the resulting nuclides with conservation of momentum. Thus, such a procedure can provide a very simple convenient way to form a hot atom excited in every possible way, translationally and electronically.

Since the first investigation of the chemical reactions of recoil [128]I atoms by Szilard and Chalmers in 1934 [2.1.13], the nuclear recoil technique has been widely used in studying the energetic reactions of various nuclides. The results have appeared in a number of reviews [2.1.14−19].

The recoil energy attained in such a nuclear reaction is far above the energy involved in chemical reactions and sometimes reaches up to 10^6 eV or more. The atoms lose most of their excess energy in collisional deexcitation processes. After continuing collisional energy losses, the atom will eventually form a chemical combination with the colliding partner in the reaction energy range. Otherwise, the atom will reach thermal energies and reacts as a thermal atom. Thus, the recoil method is characterized by the continuous energy distribution of hot atoms available for reactions. With the lack of direct control over the reaction energy, the information obtained is limited to the integration of the reactions occurred in the entire reaction energy range.

In order to use nuclear transformations as sources of hot atoms, however, the following factors must be taken into consideration:

1. Energy distribution of hot atoms in the reaction energy range. In particle emission reactions, such as (γ, n), $(n, 2n)$, etc., the recoil atom carries away a large excess kinetic energy, and a number of collisions are required for the atoms to reach the reaction energy range. This means that an equilibrium energy distribution of the atoms will be attained in the reaction range. However, the recoil energy available in radiative thermal neutron capture (n, γ) is relatively small, typically a few hundreds of eV or less: the number of collisions required to reach the reaction energy range are limited, and the equilibrium energy distribution of the atoms is attained with less certainty.

2. Charge state of the atom in the reaction energy range. Most nuclear reactions which provide enough recoil energy are more or less likely to yield the hot species as an ion. For species with keV or higher recoil energy, many charge exchange collisions occur before the species reaches the reaction energy range, so that the initial charge of the species will be usually irrelevant and the hot species will be an atom in its ground or a low-lying excited state in nearly all cases. An indication of such possibility can be afforded by the adiabatic principle [2.1.14], which provides an estimation of the relative velocity V_{max}, of the species

for maximum cross section of various types of charge exchange processes characterized by an exchange reaction energy change ΔE,

$$V_{max} = \frac{|\Delta E|\, a}{h}$$

where a is a dimension of the order of 7×10^{-8} cm. The charge exchange reaction having a smallest V_{max} principally determines the final charge state of the hot species: complete neutralization yielding a neutral atom is generally the case. Thus, for a species produced by a particle emission process, the species will be neutral in the reaction energy range. However, for the species with much less recoil energies, the adiabatic principle is no longer applicable. At least fractions of the species will be cooled down in charged states, and the ion-molecule reactions caused by these thermal ionic species become important.

3. Radiation-induced reactions. The energy deposited during bombardment of the reaction mixture with an appropriate radiation is simply proportional to the irradiation time. Thus, as the irradiation proceeds, radiation-induced reactions of both substrates and reaction products will participate in the observed product distribution. This indicates that reduction of the irradiation time is an essential requirement in conducting experiments.

4. Half-lives of recoil atoms should also be taken into account. For an atom with a longer half-life, more recoil atoms must be produced than for a short-lived atom. This will affect, in turn, the total energy deposited in the reaction mixture.

Table 2.1.2. Typical radioisotopes and nuclear processes used in hot atom studies (taken from Ref. 2.1.16)

Recoil nucleus	Half-life	Decay mode	Nuclear reaction (n* = fast neutron)
^{3}H	12.3 y	β^-	^{3}He(n, p)^{3}H ^{6}Li(n, α)^{3}H
^{11}C	20 m	β^+	^{12}C(n*, 2n)^{11}C etc.
^{14}C	5400 y	β^-	^{14}N(n, p)^{14}C
^{13}N	10 m	β^+	^{12}C(d, n)^{13}N
^{15}O	2 m	β^+	^{14}N(d, p)^{15}O
^{18}F	112 m	β^+	^{19}F(n*, 2n)^{18}F etc.
^{31}Si	2.6 h	β^-	^{30}Si(n, γ)^{31}Si ^{31}P(n*, p)^{31}Si
^{32}P	14.3 d	β^-	^{31}P(n, γ)^{32}P ^{32}S(n, p)^{32}P
^{35}S	86.7 d	β^-	^{34}S(n, γ)^{35}S ^{35}Cl(n, p)^{35}S
^{36}Cl	3×10^5 y	β^-	^{35}Cl(n, γ)^{36}Cl
^{38}Cl	37.3 m	β^-	^{37}Cl(n, γ)^{38}Cl
^{80}Br; ^{80m}Br ^{82}Br, ^{82m}Br ^{128}I	\{ Thermal neutrons initiate complex nuclear reactions with ^{79}Br, ^{81}Br and ^{127}I which give rise to both ground state and metastable excited nuclei.		

Table 2.1.2 lists some examples of radioisotopes and nuclear processes commonly used, in hot-atom studies.

Radiolysis

The production of hot hydrogen atoms through radiolysis of various organic materials is complicated by the intervention of other reactive species. Hence, the radiolysis has never been used as a tool to generate hot hydrogen atoms, while their reactions are very interesting from the viewpoint of clarification of the reaction mechanisms involved.

Hydrogen formation in the radiolysis of various hydrocarbons can be classified into two types of processes, i.e. unimolecular and bimolecular processes. The differentiation between these two processes is usually carried out by the addition of radical and electron scavengers to the system: the yield from the latter processes is independent of such additives. Later, an evidence has been presented for the important role of "hot" hydrogen atoms, that is, bimolecular hydrogen formation in the presence of radical and electron scavenger has been interpreted in terms of hydrogen atom abstraction by a hot hydrogen atom formed by direct excitation [2.1.20−23].

A rather straightforward information on the role of hot hydrogen is provided through the radiolysis of unsaturated hydrocarbons [2.1.24], which themselves act as a scavenger. The results are summarized in Table 2.1.3, together with the G-values of total hydrogen and of hot hydrogen atoms, i.e. the number of hydrogen molecules and atoms formed per 100 eV energy absorbed in the systems. The bimolecular formation is less important for ethylene but becomes more important with increasing length of the carbon chain.

Platzman [2.1.25, 26] first treated theoretically a possible role of neutral fragments in radiation chemistry by the introduction of optical approximation. According to his theory, a polyatomic molecule can be brought to a highly excited state lying above the ionization potential (superexcited state) with a certain probability. Then the hydrogen may be formed via molecular detachment from the molecules in such states, RH_2^*, or via hydrogen abstraction reaction by a hot hydrogen atom, H^*. A part of the RH_2^* may undergo autoionization. The yield of the primary decomposition channel per 100 eV of energy absorbed, g_s, may be taken as proportional to an effective dipole-matrix-element squared

Table 2.1.3. Percentages of unimolecular (denoted by I) and bimolecular (II) processes for hydrogen formation in the radiolysis of liquid olefins (taken from Ref. 2.1.24)

Olefin	$G(H_2)$	I (%)	II (%)	$G(H')^a$
trans-C_4H_8-2	1.22	18	82	1.0
C_4H_8-1	0.74	22	78	0.6
C_3H_6	0.82	29	71	0.6
C_2H_4	1.24	71	29	0.3

a H' means the precursor for bimolecular hydrogen formation in the radiolysis of liquid olefins, probably a hot hydrogen atom.

for s in atomic units:

$$M_s^2 = \int_{J_s}^{\infty} \varphi_s(E) \frac{R}{E} \frac{df}{dE} dE \tag{2.1.2}$$

where E is the excitation energy (in eV), R the Rydberg energy, df/dE the differential oscillator strength, $\varphi_s(E)$ the probability of decomposition upon excitation at E, and J_s the threshold excitation energy value (in eV). With M_i^2, the dipole-matrix-element squared for ionization,

$$M_i^2 = \int_I^{\infty} \eta(E) \frac{R}{E} \frac{df}{dE} dE, \tag{2.1.3}$$

the g_s can be expressed by

$$g_s = (100/W) \cdot (M_s^2/M_i^2). \tag{2.1.4}$$

In the numerical calculation of Eqs. (2.1.3) and (2.1.4), the data on photo-absorption and ionization are utilized. The results are still very tentative because of the approximate nature of the procedure, but will amply indicate the significant role of superexcited states of hydrocarbons in radiolysis.

References

[2.1.1]   Siska, P. E., Parson, J. M., Schafer, T. P., Lee, Y. T.: J. Chem. Phys. 55, 5762 (1971)
[2.1.2]   Shobatake, K., Lee, Y. T., Rice, S. A.: ibid. 59, 6104 (1973)
[2.1.3]   Herm, R. R., Lin, S.-M., Mims, C. A.: J. Phys. Chem. 77, 2931 (1973)
[2.1.4]   Fite, W. L., Datz, S.: Ann. Rev. Phys. Chem. 14, 61 (1963)
[2.1.5]   Beatty, J. W., Wexler, S.: J. Phys. Chem. 75, 2417 (1971)
[2.1.6]   Beatty, J. W., Pobo, L. G., Wexler, S.: ibid. 75, 2407 (1971)
[2.1.7]   Menzinger, M., Wolfgang, R.: J. Chem. Phys. 50, 2991 (1969)
[2.1.8]   Leroy, R. L., Yencha, A. J., Menzinger, M., Wolfgang, R.: ibid. 58, 1741 (1973)
[2.1.9]   Williams, R. R. Jr., Odd, R. A. Jr.: ibid. 15, 676 (1947)
[2.1.10] Martin, R. M., Willard, J. E.: ibid. 40, 3007 (1964)
[2.1.11] Chou, C. C.: Ph. D. Thesis, Univ. of California, Irvine, (1968)
[2.1.12] Fink, R. D., Nicholas, J. E.: J. Chem. Soc., Faraday Trans. 68, 1706 (1972)
[2.1.13] Szilard, L., Chalmers, T. A.: Nature (London) 134, 462 (1934)
[2.1.14] Wolfgang, R.: Progr. Reaction Kinetics 3, 97 (1965)
[2.1.15] Urch, D. S.: MTP. Int. Rev. Sci., Inorg. Series, Two 8, 49 (1975)
[2.1.16] Urch, D. S.: ibid., Inorg. Series, One 8, 149 (1972)
[2.1.17] Wolf, A. P.: Advan. Phys. Org. Chem. 2, 202 (1964)
[2.1.18] Rowland, F. S.: MTP. Int. Rev. Sci., Phys. Chem. Series, One 9, 109 (1972)
[2.1.19] Tang, Y. N.: in Isotopes in Organic Chemistry E. Buncel, C. C. Lee (eds.) 4, 85, Elsevier 1978
[2.1.20] Manion, J. P., Burton, M.: J. Phys. Chem. 56, 560 (1952)
[2.1.21] Hardwick, T. J.: ibid. 66, 1611 (1962)
[2.1.22] Voevodskii, V. V., Molin, Y. M.: Radiat. Res. 17, 366 (1962)
[2.1.23] Hatano, Y.: Bull. Chem. Soc. Jpn. 41, 1126 (1968)
[2.1.24] Hatano, Y., Shida, S., Sato, S.: ibid. 41, 1120 (1968)
[2.1.25] Platzman, R. L.: Vortex 23, 372 (1962)
[2.1.26] Platzman, R. L.: Radiat. Res. 17, 419 (1962)

2.2 Radiochemical Separation Technique

2.2.1 Gas Phase: Radiogas Chromatography

In hot atom experiments, the reaction processes are understood through analysis of the products without disturbing the primary product distribution. The number of hot atoms involved in the experiments is always very small (usually of the order of $10^6 - 10^8$ atoms). Thus, the ordinary separation techniques commonly used for macroscopic amounts are not sensitive enough to detect the products in the hot atom experiments. Hence, the recommended methods should be based on the detection of radioactivities.

A standard analytical device for the gas (or liquid)-phase hot atom experiments is a gas chromatograph equipped with a radiation detector, called a radiogas chromatograph, where the products can be separated by gas chromatography and the yields of individual recoil products are determined by measuring their radioactivities. Meanwhile, macroscopic amounts of products can be measured by ordinary gas chromatography.

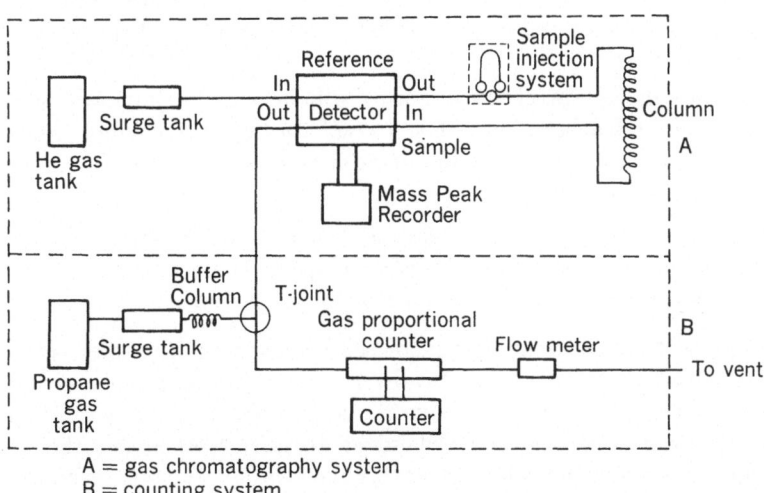

A = gas chromatography system
B = counting system

Fig. 2.2.1. Block diagram of a radiogas chromatograph (taken from Ref. 2.2.4)

Kokes et al [2.2.1] first reported the use of a GM counter as detector in series with a gas-chromatographic column. Halogen-labeled compounds were also radioassayed [2.2.2]. Wolfgang and Rowland [2.2.3] have developed a technique by combining a gas chromatograph with a gas-flow proportional counter for the radioassay of soft β-emitter nuclides. In their method, methane or propane is injected into the carrier gas stream to make up a suitable counter gas. Figure 2.2.1 shows a schematic diagram of their experimental set-up. In this figure, the upper, part is a regular gas chromatograph with a thermal conductivity detector. In the lower part, a radioactive counting device is shown. This combined system offers the following advantages over the conventional methods of separation and radioassay.

a) The radioassay can be made simultaneously with the separation and mass assay.

b) Since a complete radiochromatogram is obtained, it is not likely that any portion of the activity in an unexpected chemical form is missing.

c) Carrier-free radioactive compounds can be handled with the same technique as for macroscopic quantities.

Column Separation

Recently, various separation columns have been examined and any compound can be separated by choosing an appropriate column, or combined columns. In some cases, mutual separations of isotopic compounds are also needed: successful columns are listed in Table 2.2.1 [2.2.4]. The following considerations should be necessary in performing experiments:

a) Since the separation of the compounds is not the final goal, the flow rate and column length should be chosen to be suitable for radioactive counting. The peak separation in the column must be large enough to prevent the components from remixing in the detector and to provide sufficient resolution of the activity peaks.

b) Radioactivities in certain chemical forms will be lost by processes such as isotopic exchange and absorption, while passed through the column. Thus, any compound tagged with a radionuclide in reactive or labile positions cannot be subjected to the column separation.

For the purpose of separation of various mixtures within a limited length of time, special techniques have also been developed. These are "recycling" [2.2.5] and "double column" [2.2.6] techniques described below.

Recycling Technique. It is apparent that increased column length enables better resolution of close neighboring peaks. However, the technical problems such as flow rates and overpressure usually limit the length of the column to 15 or 30 m in most practical applications. Rowland et al. have worked out the recycling technique to extend the column length almost infinitely; this technique has proved applicable to the separation of isotopic molecules and other molecular pairs hardly separable on usual columns. The chromatographic system [2.2.5] used for this technique operates with two identical columns connected in series (C_1 and C_2 in Fig. 2.2.2) and a detector in between (R_1). The columns are connected with four-way stop-cocks (S_1 and S_2) for rapid reversal of the sequence of the columns in the series; and an additional detector (R_2) is provided at the exit of the system. The sample was injected into one of the separation column through a short stripper column. After the sample had moved into the separation column, the stripper column and injection apparatus are by-passed through. After the transfer of the components from one separation column to another, the four-way stopcocks C_1 and C_2 are switched from the position A to position B. The transfer of the components will be monitored through the detector (R_1).

Table 2.2.1. Isotopic separation of some tritiated compounds by gas chromatography (taken from Ref. 2.2.4)

T-labelled compounds	Chromatographic column	Operation temperature (°C)	Separation
HT/DT	γ-Alumina coated with 3.5% Fe_2O_3	-198	Complete
HT/DT	Molecular sieve 13 A	-160	Complete
H_2/HT/T_2	Molecular sieve	-160	Complete
CH_3T/CH_2T_2/ CHT_3/CT_4	Highly activated charcoal	$-$ 3.5	Partial
CH_3T/CD_3T	Activated charcoal	52	Partial
CH_3T/CH_2T_2/ CHT_3/CT_4	Etched soft glass capillary column	-198	Almost complete
CH_3T/CHD_2T/CH_2DT	Charcoal	50	Partial
$CHTCH_2$/C_2H_4	$AgNO_3$-ethylene glycol	0	Partial
$CHTCH_2$/$CHTCD_2$	$AgNO_3$-ethylene glycol	0	Partial
$CH_2TCH{=}CH_2$ vs $\begin{cases} CH_3{-}CT{=}CH_2 \\ CH_3{-}CH{=}CHT \end{cases}$	$AgNO_3$-ethylene glycol	0	Partial
$CD_2TCH{=}CH_2$ vs $CH_2TCH{=}CD_2$	$AgNO_3$-ethylene glycol	0	Partial
$\begin{cases} CH_2TCH_2CH{=}CH_2 \\ CH_3CHTCH{=}CH_2 \end{cases}$ vs $\begin{cases} CH_3CH_2CT{=}CH_2 \\ CH_3CH_2CH{=}CHT \end{cases}$	$AgNO_3$-ethylene glycol	0	Partial
c-C_4H_8/c-C_4H_7T	Safrole column with recycling	52	Partial

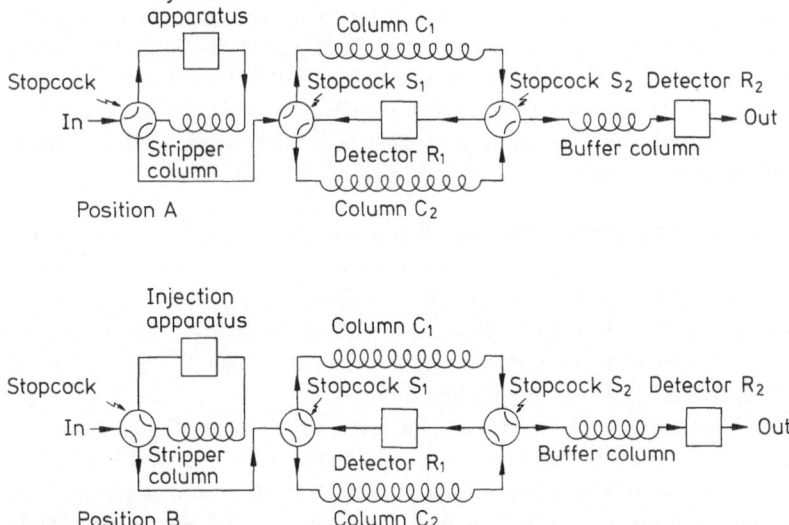

Fig. 2.2.2. Recycling set-up [2.2.6]

Thus, whenever required, an appropriate non-radioactive carrier molecules can be added to the mixed components, prior to the separation. The recycling process was carried out until the desired separation was obtained (indicated by the mass peaks) or until the peaks had spread out to the whole column length and further recycling was not feasible. This "tail-eating" problem limits the maximum recycling period of a given experimental set-up.

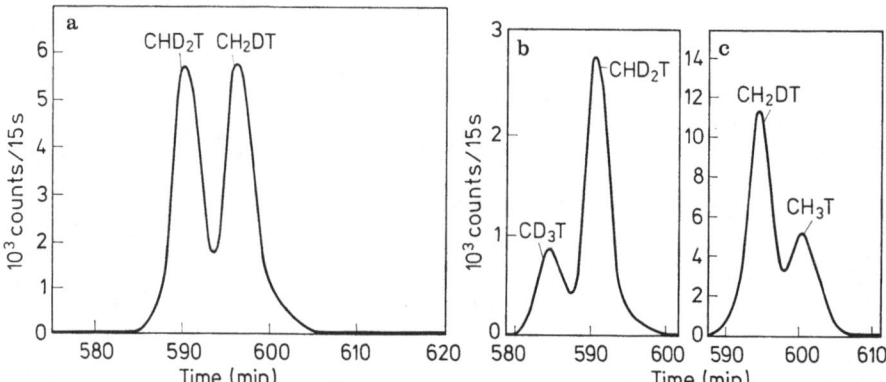

Fig. 2.2.3. Isotopic separation of pairs of (a) CH_2T-CH_2DT, (b) CD_3T-CHD_2T, and (c) CH_2DT-CH_3T [2.2.7]. Column, 105-m Graphon + 0.1% squalene; temperature, $-105\,°C$; inlet pressure, 15 kg/cm^2; flow rate, 320 ml/min

Analytical separation of isotopic methanes was performed on two identical columns, 105m columns of GRAPHON with 0.1% squalene (Fig. 2.2.3) [2.2.7], or 30.5m columns of charcoal [2.2.8]. The efficiency of the isotopic separation is a function of the differences in the number of H vs. D atoms in the molecules to be separated. In general, the separation of pairs of methanes differing by two or three H/D changes (e.g., CH_3T vs. CHD_2T or CD_3T) is very satisfactory. Separation for isotopic methanes has been improved with capillary columns and relatively small-size sample injections [2.2.9, 10].

Double Column Technique. When the sample gas consists of various components from low to high molecular weights, a single column cannot provide satisfactory separation of each component. The double column technique is often suitable, particularly when the sample cannot be devided without difficulty. This technique can achieve total resolution of mixtures: otherwise, it is usually necessary to analyze at least two aliquots on separated columns.

The system is essentially the same as the recycling system (Fig. 2.2.2), but with two different columns. The sample is injected into the first column, C_1. Immediately after the mixed components which have not been separated in columns C_1 enter column C_2, both four-way stopcocks S_1 and S_2 are switched. The mixed components are separated subsequently in column C_2, are then repassed through column C_1, and finally enter the end cell. Those compounds with longer retention times on column C_1 than the mixed components will pass directly through column C_1 into the final cell. The buffer column is needed between the stopcock S_2 and the final cell R_2, in order to minimize the pressure effects of switching. A typical radiogas chromatogram is shown in Fig. 2.2.4 [2.2.6].

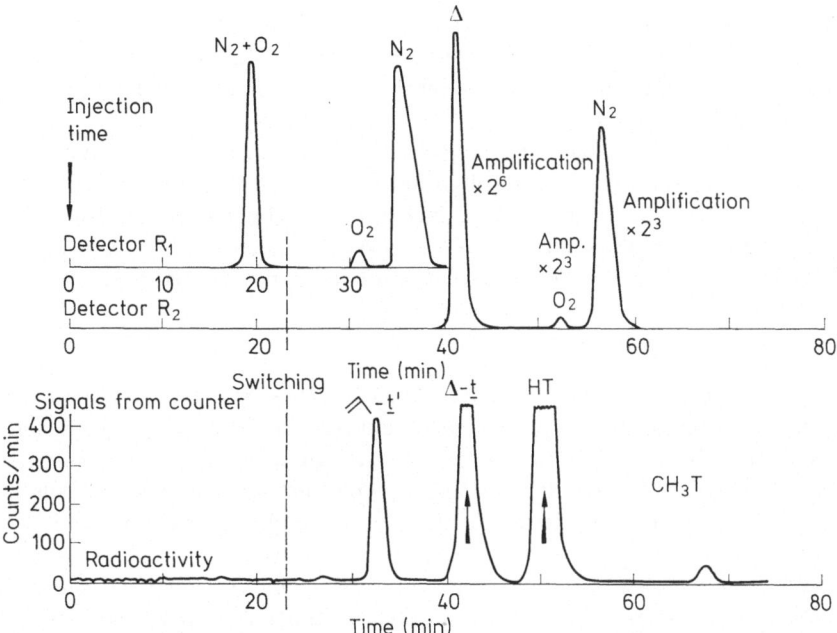

Fig. 2.2.4. Cycled chromatography [2.2.6]. Cyclopropane-N_2 system (column *1*: DMS; column *2*: MS-10 X)

Radioactivity Assay

In radiogas chromatography, an adequate counting device should be selected for the particular radionuclide under study. For hard γ or β emitters, a conventional scintillation or GM counter is appropriate. In these cases, the carrier gas stream is simply allowed to flow past a thin-walled counter. Certain radionuclides emit so weak radiations that, in the case of tritium, for example, they cannot penetrate the counter wall at all. This problem could be overcome by passing the gas through inside the counter. However, helium and nitrogen, preferred carrier gases in gas chromatography, make poor counter-filling gases. An appropriate gas is added to the effluent of the gas chromatograph, so as to make up a counting gas [2.2.3, 11]. With this technique, most proportional counters can be used with a given gas mixture.

The counter size and the flow rate must be selected in accordance with the relative choice of increased resolution and increased sensitivity. The limit of sensitivity will be determined by both the total counts recorded (N) and the average counting rate (R) due to a given peak, which are given by [2.2.11]

$$N = A[v/(f_c + f_a)], \tag{2.2.1}$$

$$R = (N/CV) (f_c + f_a) = Av/CV \tag{2.2.2}$$

in which A is the total absolute activity of material in this peak (dinsintegrations per second), v the counter volume (ml), f_c and f_a are flow rates of carrier gas and added counting gas, respectively (ml s^{-1}), V the retention volume (ml), and C a dimensionless constant. The fractional standard deviation of the result is approximately $[(1/N) (1 + R_b/R)]^{1/2}$, where R_b is the background counting rate. Precision is thus improved either by increasing N or by making R large relative to R_b.

Two of the variables which can be adjusted in a given run are the flow rate and, by varying the temperature, the retention volume. For operation at maximum sensitivity, low flow rates are used to increase the total counts recorded (N). In addition, the average counting rate (R) may be increased by operating at the highest temperature permitting resolution of the narrow initial peak.

Choice of Counting Gas Mixture

When the same gas serves adequately both as the flow gas in chromatography and as the counting gas in the proportional counter, the counting considerations are greatly simplified. This is, however, not met in most cases. If a thermal conductivity detector is used for the macroscopic measurements, as is usually the case, helium is most preferred because of its high thermal conductivity. With this carrier gas, proper counter operation requires the addition of an adequate amount of a gas to make an appropriate mixture for gas proportional counter. While many organic gases will provide such suitable mixtures, methane or propane is frequently used and gives a satisfactory performance of the counter: propane is generally preferable to methane but helium-methane mixtures can be very useful in experiments requiring trapping of some volatile components from this stream.

Variations in Detection Efficiency

The most common causes for variations in counting efficiency of a gas mixture in a counter are coincidence losses at high counting rates and macroscopic changes in the gas composition during the passage of separated components through the counter. When a counter was operated on filling the gases, rather than under flow conditions, a very similar response was obtained to an internal tritium filling and to external radium and ^{22}Na sources [2.2.12]. Thus, the characteristics of a given counter can be conveniently examined using an external radioactive source. With helium-methane mixtures, higher CH_4/He ratios expand the plateau and at the same time raise both the threshold and the plateau to higher voltages (Fig. 2.2.5). At higher counting rates, a flat plateau obtained with low counting rates is no longer obtained. The resolving time for these gas mixtures increases markedly with the increase in the applied voltage, resulting in heavy coincidence loss at higher voltages and a dip in the observed count rates. The propane-helium mixture exhibits similar characteristics, but the falloff in count rate-vs.-voltage is less severe. Thus, the coincidence loss can be minimized by operating the counter near the beginning of the plateau in the region with shortest resolving time.

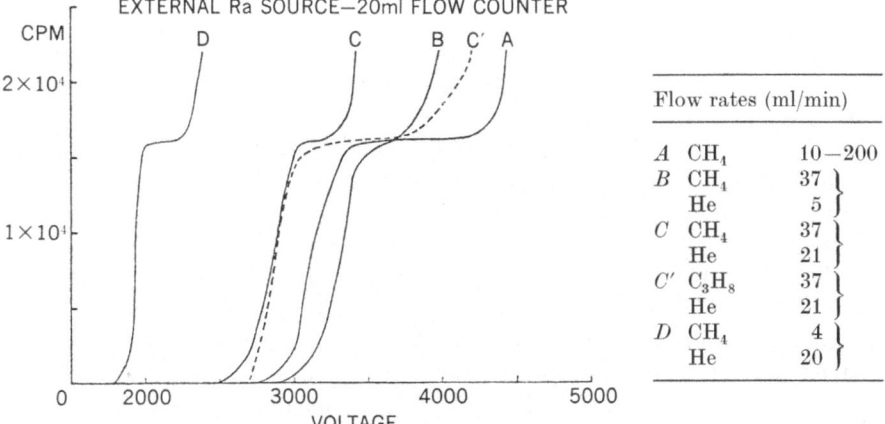

Fig. 2.2.5. Count rate vs. voltage plateaus for CH_4—He mixtures. (Reprinted with permission from Ref. 2.2.11. Copyright (1962) American Chemical Society)

Except for certain compounds, none of the gases have a great effect on the counter operation despite the large quantities of the gas involved. However, when a compound with a high electron affinity passes through the counter, counting efficiency is reduced due to the electron attachment processes. The change in counting efficiency is not necessarily linear with the amount of these gases, and must be determined experimentally beforehand. The composition change in the counter during the passage of a large macroscopic peak will raise the plateau to a higher voltage, and the efficiency will be lowered. Operation at voltages well above the "knee" of the plateau will be preferred; the instantaneous shift of the plateau above this voltage will be less likely to happen.

2.2.2 Condensed Phase

Various conventional separation techniques involving wet chemistry have been employed to separate radioactive products resulting from the recoil reactions in condensed phase. Of these, chromatographic means such as ion exchange, paper electrophoresis and column chromatography appear most useful for systematic separation and analysis of numerous recoil species to obtain an overall picture of their identities and radiochemical yields. While solvent extraction and precipitation are convenient for handling of relatively large-sized samples, they can only separate the fraction containing a particular species from the rest of the sample, but never provide complete information regarding the chemical distribution of all the recoil products. Chromatographic separation technique based on sublimation of solid samples also proves to be a promising means for analysis of the solid systems since no dissolution process (i.e. wet chemistry) is involved in this technique. Special separation technique utilizing positive charges or kinetic energies of the recoil species will be mentioned later in Sect. 4.1.

Ion Exchange. Ion exchange separation has been commonly used for analysis of the recoil species produced in complex salts or oxyanion salts. Radioactive cations or anions are adsorbed on a cation or anion exchanger column and eluted chromatographically with various eluants (aqueous solutions of acids, salts or chelating agents). Figure 2.2.6 indicates a typical ion exchange chromatogram of irradiated $[Co(NH_3)_6](NO_2)_3$ [2.2.13]. Cis- and trans-isomers can also be separated successfully by ion exchange method [2.2.14]. For simpler separation of enriched radioisotopes, ion exchange resins or inorganic ion exchangers loaded with target (complex) ions were irradiated and then the radionuclides (^{64}Cu, ^{60}Co, ^{56}Mn, ^{76}As, etc.) released by recoil were eluted out with high enrichment factors [2.2.15—17]. A dynamic separation method has been proposed for ion exchange separation of metal complexes: the complex ion adsorbed on an ion exchange column is irradiated in a reactor while an eluant is passed through to elute the product species [2.2.18, 19]. Higher separable yields and specific activities can be expected with this technique since the secondary reactions (e.g. recombination) are suppressed by continuous elution.

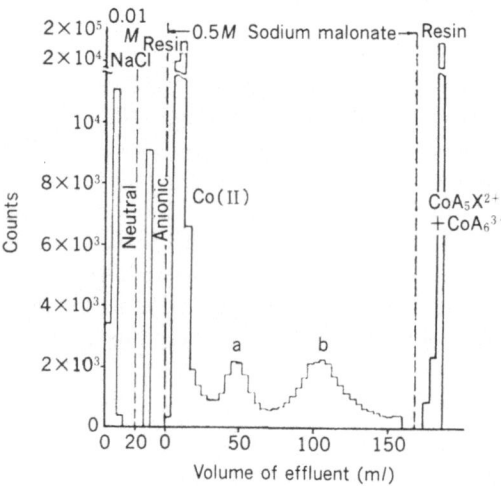

Fig. 2.2.6. Ion exchange chromatogram of ^{60}Co recoil species in irradiated $[Co(NH_3)_6](NO_2)_3$ (from Ref. 2.2.13). Cation exchange column: Diaion SK # 1 (8 mm $\varnothing \times 200$ mm). Anion exchange column: Dowex 1-X8 (10 mm $\varnothing \times 200$ mm). *a*, trans-$CoA_4X_2^+$; *b*, cis-$CoA_4X_2^+$

Paper Electrophoresis. Although only a small amount of sample can be taken for analysis, paper electrophoresis has provided an excellent means for separation of recoil products in metal complexes and phosphates [2.2.20—22]. On application of a high voltage, the ionic species present in the solution contained in a paper strip move towards each electrode according to their charges. Figure 2.2.7 illustrates an electrochromatogram of neutron-irradiated $Na_2IrCl_6 \cdot 6H_2O$ [2.2.22, 23]. At least thirteen ^{192}Ir-labeled recoil species can be identified in the chromatogram.

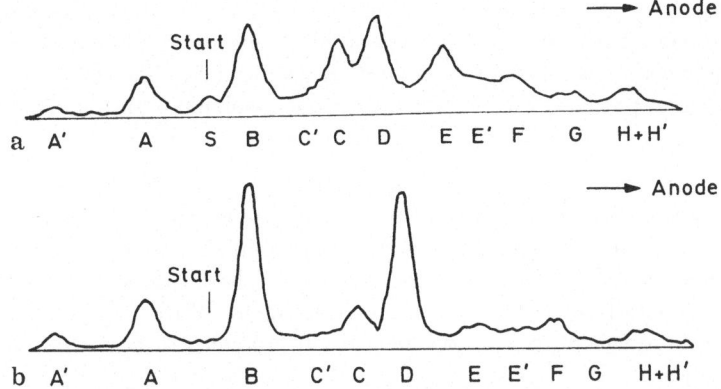

Fig. 2.2.7. Electrochromatograms of a solution of neutron-irradiated $Na_2IrCl_6 \cdot 6H_2O$, (**a**) freshly prepared, and (**b**) after 6 days (taken from Ref. 2.2.22)

A': possibly $[IrCl(H_2O)_5]^{2+}$
A: possibly $[IrCl_2(H_2O)_4]^+$
S: IrO_2 and/or Ir_2O_3
B: $IrCl_3(H_2O)_3$
C': very likely $[IrCl_3(H_2O)_2OH]^-$
C: very likely $[IrCl_3(H_2O)(OH)_2]^{2-}$
D: $[IrCl_4(H_2O)_2]^-$

E: very likely $[IrCl_4(H_2O)OH]^{2-}$
E': very likely $[IrCl_4(OH)_2]^{3-}$
F: $[IrCl_5(H_2O)]^{2-}$
G: probably $[IrCl_5OH]^{3-}$
H: $[IrCl_6]^{2-}$
H': $[IrCl_6]^{3-}$

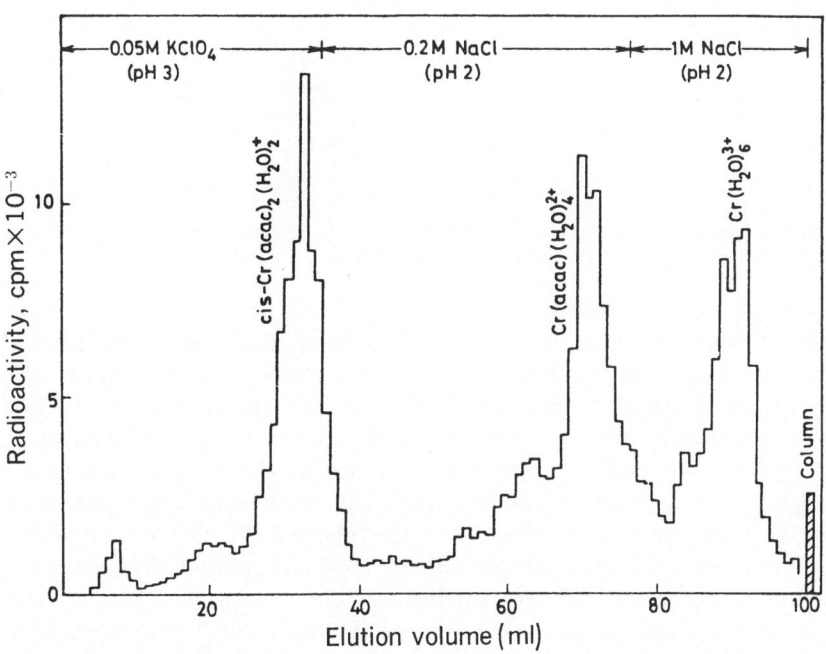

Fig. 2.2.8. Chromatogram of ^{51}Cr-labeled species extracted into aqueous phase from irradiated $Cr(acac)_3$ (from Ref. 2.2.24). (SP Sephadex C-25 column: K-form, 10 mm $\varnothing$ $\times$ 110 mm)

Column Chromatography. Recoil products formed in organometallic molecules or neutral chelate complexes can be often separated by means of column chromatography. The species absorbed on a column (e.g. alumina, silica gel, etc.) are eluted chromatographically with non-polar or polar solvents. In Fig. 2.2.8 is shown a chromatogram of irradiated $Cr(acac)_3$ [2.2.24].

Solvent Extraction. Solvent extraction has been applied to separate 'free' recoil atoms from complex compounds. These 'free' recoil atoms can be extracted (enriched) from the organic solution of the target compound into aqueous phase. The 'free' atoms arising from the complex ions in aqueous phase can be extracted into organic phase containing an adequate chelating agent. With a particular extraction process, however, only one (or a few) recoil product can be separated from the other species.

Precipitation. Precipitation or coprecipitation is now seldom used in analysis of recoil products except for metal phthalocyanines, for example. The precipitation technique has been successfully applied to separate two similar recoil species from each other, which cannot be separated by chromatographic means [2.2.25]. For instance, $^{60}Co(NH_3)_5NO_2^{2+}$ and $^{60}Co(NH_3)_5NO_3^{2+}$ were formed in irradiated $[Co(NH_3)_5NO_2](NO_3)_2$, and the yield of each complex was determined by repeating recrystallization to constant specific activity.

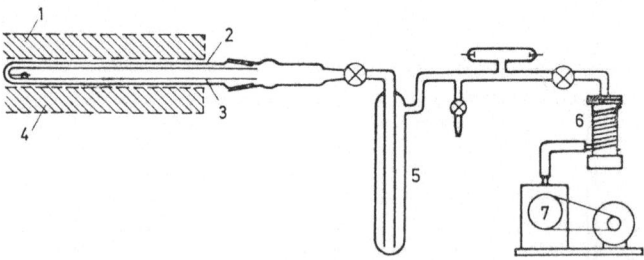

Fig. 2.2.9. Sublimation apparatus (from Ref. 2.2.27). *1* Electric furnace. *2* Outer quartz tube. *3* Inner glass tube. *4* Sample boat. *5* Cold trap. *6* Oil diffusion pump. *7* Rotary pump

Sublimation. Vacuum sublimation technique can be applied to the radiochemical separation of volatile organometallic compounds. Metal carbonyls containing fission-produced nuclei have been separated from the irradiated solid uranium oxide and non-volatile fission products [2.2.26]. Sakanoue and his coworkers has developed a sublimation apparatus as shown in Fig. 2.2.9, in which the volatile species are deposited chromatographically within an inner glass tube according to the temperature gradient along the tube [2.2.27]. The recoil products from the irradiated solid metal acetylacetonates and phthalocyanines have been separated by means of the sublimation chromatography [2.2.27—29]. Although the recoil species inevitably undergo thermal annealing during heating process for sublimation of the samples, the sublimation technique may be useful for qualitative study of the recoil species since the irradiated solid systems can be analyzed without dissolution. For instance, formation of $^{60}Co(II)$ recoil species (probably ^{60}Co(acac)$_2$) in the irradiated solid Co(acac)$_3$ has been assumed by comparing the reactions after dissolution in air and in argon (wet chemistry) [2.2.30]. The formation of ^{60}Co(acac)$_2$ in the irradiated solid Co(acac)$_3$ has been verified later by sublimation chromatography as revealed in Fig. 2.2.10 [2.2.31].

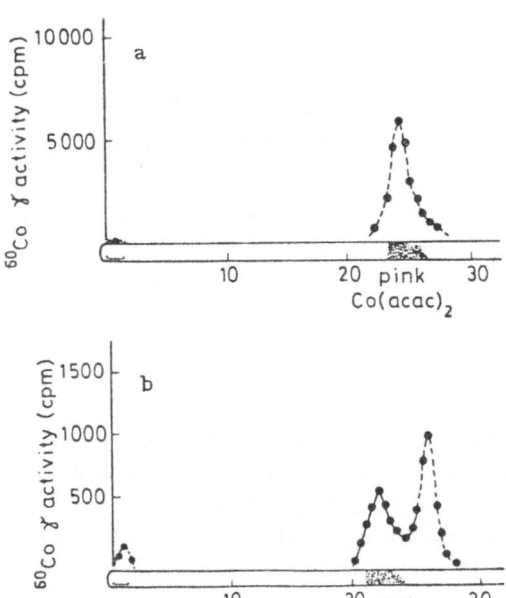

Fig. 2.2.10. Sublimation chromatograms of (a) irradiated Co(acac)$_2$, and (b) irradiated Co(acac)$_3$ (from Ref. 2.2.31)

References

[2.2.1]   Kokes, R. J., Tobin, H. Jr., Emmett, P. H.: J. Am. Chem. Soc. 77, 5860 (1955)
[2.2.2]   Evans, J. B., Willard, J. E.: ibid. 78, 2908 (1956)
[2.2.3]   Wolfgang, R., Rowland, F. S.: Anal. Chem. 30, 903 (1958)
[2.2.4]   Tang, Y. N.: in Isotopes in Organic Chemistry E. Buncel, C. C. Lee (eds.) 4, 85, Elsevier 1979
[2.2.5]   Root, J. W., Lee, E. K. C., Rowland, F. S.: Science 143, 676 (1964)
[2.2.6]   Tachikawa, E.: Ph. D. Thesis. University of California, Irvine, 1964
[2.2.7]   Chou, C. C., Rowland, F. S.: unpublished results
[2.2.8]   Chou, C. C., Rowland, F. S.: J. Phys. Chem. 76, 1283 (1971)
[2.2.9]   Bruner, F., Cartoni, G. P.: J. Chromatogr. 18, 390 (1965)
[2.2.10] Bruner, F., Cartoni, G. P., Liberti, A.: Anal. Chem. 41, 1122 (1969)
[2.2.11] Lee, J. K., Lee, E. K. C., Musgrave, B., Tang, Y. N., Root, J. W., Rowland, F. S.: Anal. Chem. 34, 741 (1962)
[2.2.12] Rowland, F. S., Lee, J. K., White, R.: Oklahoma Conference, TID-7578, USAEC. p. 39, 1959
[2.2.13] Ambe, S., Sano, J., Tominaga, T., Saito, N.: Radioisotopes 21, 543 (1972)
[2.2.14] Saito, N., Sano, H., Tominaga, T., Ambe, F., Fujino, T.: Bull. Chem. Soc. Jpn. 35, 744 (1962)
[2.2.15] Duncan, J. F.: Radioisotope Techniques II, HMSO, p. 121, London 1952
[2.2.16] Buser, W., Graf, P., Imoberstag, U.: Z. Elektrochem. 58, 605 (1954)
[2.2.17] Saito, N., Furukawa, M., Tomita, I.: J. Chem. Phys. 27, 1432 (1957)

[2.2.18] Sensui, Y., Matsuura, T.: Bull. Chem. Soc. Jpn. *38*, 1171 (1965)
[2.2.19] Matsuura, T., Sensui, Y., Sasaki, T.: Radiochim. Acta *4*, 85 (1965)
[2.2.20] Saito, N., Tominaga, T., Sano, H.: J. Inorg. Nucl. Chem. *24*, 1539 (1962)
[2.2.21] Lindner, L., Harbottle, G.: J. Inorg. Nucl. Chem. *15*, 386 (1960)
[2.2.22] Bell, R., Herr, W.: Radiochim. Acta *2*, 125 (1964)
[2.2.23] Herr, W., Heine, K., Schmidt, G.: Z. Naturforsch. *17a*, 590 (1962)
[2.2.24] Omori, T., Shiokawa, T.: Radiochim. Radioanal. Lett. *17*, 167 (1974)
[2.2.25] Saito, N., Tominaga, T., Sano, H.: Bull. Chem. Soc. Jpn. *36*, 232 (1963)
[2.2.26] Kienle, P., Baumgärtner, F., Zahn, U.: Radiochim. Acta *1*, 84 (1963)
[2.2.27] Kawazu, H., Sakanoue, M.: Radiochem. Radioanal. Lett. *16*, 373 (1974)
[2.2.28] Amano, R., Sakanoue, M.: Radiochem. Radioanal. Lett. *16*, 381 (1974)
[2.2.29] Endo, K., Sakanoue, M.: Radiochem. Radioanal. Lett. *9*, 255 (1972)
[2.2.30] Tominaga, T., Nishi, Y., Motohashi, E.: Radiochem. Radioanal. Lett. *18*, 15 (1974)
[2.2.31] Amano, R., Sakanoue, M.: Radiochem. Radioanal. Lett. *19*, 197 [(1974)

2.3 Special Physical Techniques

2.3.1 Special Physical Technique in Gas Phase Studies — Charge Spectrometry

Common experimental procedures are based on the identification of stable products produced by the reactions of recoil species with reactant molecules. Thus, the mechanisms and kinetics can be deduced only in accordance with the observed product distributions. In order to verify the proposed mechanisms and acquire further understanding of the reactions involved, the following information will be required [2.3.1]:

a) The properties of the recoil species immediately after its birth in the nuclear process.

b) The modes of translation- and excitation-energy loss and the change of charge during slowing down of the energetic recoils into the range of chemical-reaction energies.

c) The identity of the intermediate species participating in the possible chemical reactions leading to the observed products.

The properties of the recoil species can only be investigated on the recoils from radioactive decay because of the sensitivity for its detection. Two processes mainly take place in the electron shells of an atom following a radioactive decay in the nucleus [2.3.2]. The first process is the creation of vacancy cascades. If the nuclear transformation is of such a nature as to remove an extranuclear electron from a deep-lying shell (i.e. the K- or L-shell), there exists a good possibility that the vacancy creation will be followed by an Auger event, leaving a multiply-charged atom (or molecule). This is usually the case with the emission of a conversion electron in the isomeric transition. The second is the so-called electron shake-off process. When a nucleus emits a charged particle, the nuclear charge is altered correspondingly. Usually the particle escapes from the atom so quickly that an abrupt perturbation in the electrostatic environment is felt by all extranuclear electrons, possibly leading to electronic excitation, or even ionization. Although the probability for vacancy creation is proportional to

$1/Z_{\text{eff}}$, and the electrons in outer shells have higher probability for ionization, a deep-lying electron is also shaken off on rare occasions and a vacancy cascade is likely triggered.

The experimental investigations of these phenomena have been successfully performed by employing a charge-spectrometer, in which the charge of product ions can be measured before they have a chance to make charge-exchange collisions with other atoms or molecules, and the relative probabilities for the loss of one, two, or more electrons as the consequence of the radioactive decay are determined rather directly. Figure 2.3.1 is the schematic diagram of a charge-spectrometer employed by Wexler [2.3.3]. The gaseous radioactive substance is introduced through a capillary leak into the conical source volume (80 l), where it decays under a vacuum in the range $(3—14) \times 10^{-6}$ mmHg. The positive ions born in the decay are attracted toward the apex of the cone by an electric field if proper potentials are applied to the series of guide rings. The emerging ions pass through an adjustable slit and the mass analysis is performed by deflecting them magnetically.

Snell and Pleasonton have measured the relative abundances of ions resulting from the internal conversion and electron capture in various rare gases [2.3.2]. It was found that 11% of the decayed ^{85}Kr underwent shake-off of 4s or 4p

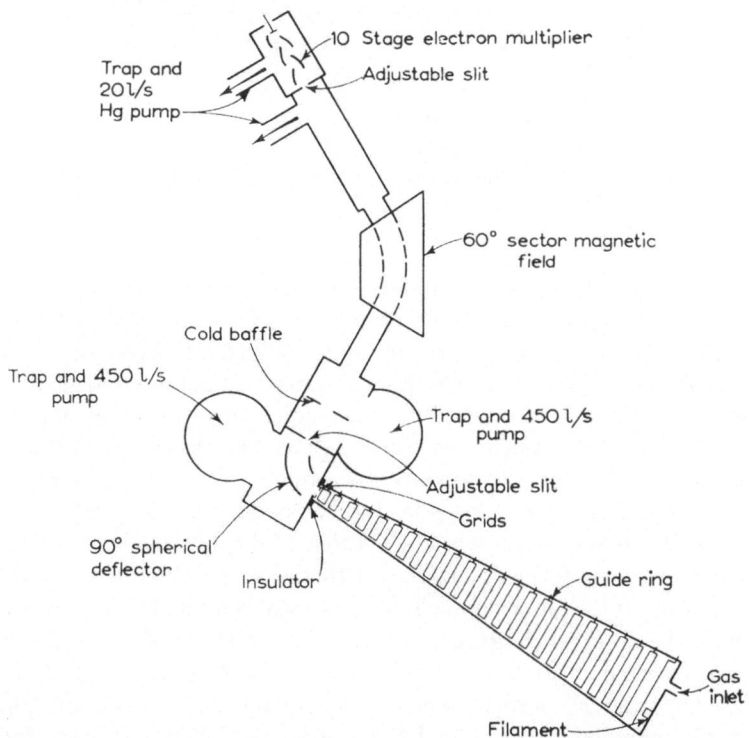

Fig. 2.3.1. Mass spectrometer for radioactive gases (taken from Ref. 2.3.3)

electron, and further 10% suffered from self-ionization in deeper shells. In the isomeric transition of ^{131}Xe, 21% of the internal conversion events lead to removal of 8 electrons, while only one conversion electron is emitted in 0.62% of the events and 20 electrons are ejected in 0.003% of the events. The measurements were further extended to molecular systems by other research groups [2.3.4, 5]. In the decay of tritium in HT [2.3.4], the $(^3HeH)^+$ ions constitute about 90% of the total positive ions observed while, in the various tritiated hydrocarbons, C-^{3}He bond is almost always ruptured and the positive charge increment on the ^{3}He atom is transferred onto the organic fragment. The organic fragment may be further dissociated through C-C bond rupture or loss of neutral H and H_2.

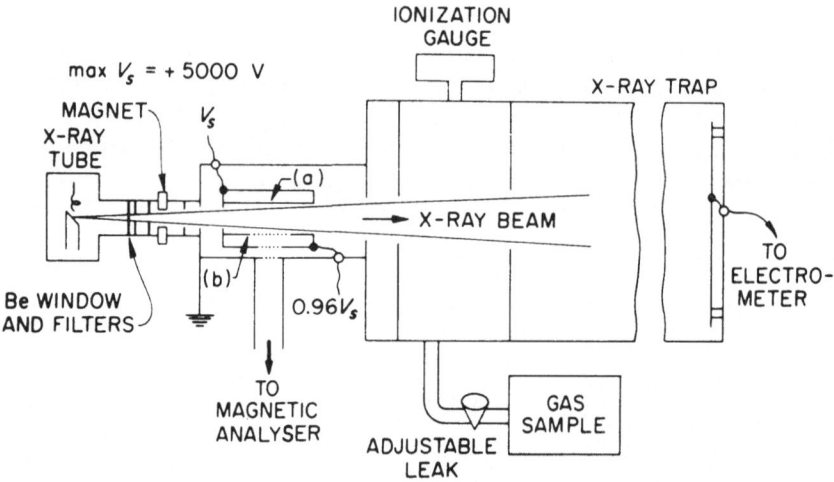

Fig. 2.3.2. Source volume for studying the charge distribution of ions produced by X-rays (taken from Ref. 2.3.6)

The initial vacancy in an inner shell of an atom, initiating the vacancy cascade, was also created by means of X-ray irradiation by Carlson and White [2.3.6]. The experimental set-up was very similar to that in Fig. 2.3.1, except for the source volume as shown in Fig. 2.3.2. The X-ray source was operated at 40 keV with a tungsten target. The advantage of this technique consists in that experiments are not limited at all to radioactive isotopes alone, and that a proper choice of X-ray energy makes it possible to produce a vacancy in most of the shells.

When CH_3I is used as a target molecule, the X-ray source will produce an initial vacancy mostly in the L shell of iodine. Table 2.3.1 lists the relative abundances of the fragment ions formed, together with the peak values of the recoil energy spectra for each of the ions examined. Very low yields for molecular ions are noticed. Nearly all the observed fragments are H^+, C^{n+}, or I^{n+}, in sharp contrast to the data from electron impact, or decomposition following β^--decay. Since the sums of I^{n+}, C^{n+}, and H^+ ions respectively are in the ratio of 1.0:1.0:3.0, indicating that all component atoms in CH_3I are ionized, they proposed the multi-ion Coulomb explosion model: as the consequence of the vacancy cascade

Table 2.3.1. Relative abundances and recoil energies of the fragment ions from the decom position of CH_3I following an inner shell vacancy in iodine (from Ref. 2.3.6)

Ion	Relative abundance (Relative to I^{5+} = 1.00)	Recoil energy[a] (eV)
I^{1+}	0.20 $\pm$ 0.02	—
I^{2+}	0.42 $\pm$ 0.02	2.2 $\pm$ 0.4
I^{3+}	0.59 $\pm$ 0.02	4.5 $\pm$ 0.8
I^{4+}	0.82 $\pm$ 0.02	5.8 $\pm$ 1.0
I^{5+}	1.00	8.6 $\pm$ 1.0
I^{6+}	0.62 $\pm$ 0.02	10 $\pm$ 2
I^{7+}	0.50 $\pm$ 0.02	14 $\pm$ 3
I^{8+}	0.24 $\pm$ 0.01	18 $\pm$ 3
I^{9+}, CH_2^+	0.10 $\pm$ 0.01	25 $\pm$ 6
I^{10+}	0.03 $\pm$ 0.01	—
I^{11+}	0.007 $\pm$ 0.003	—
ΣI^{n+}	4.53 $\pm$ 0.05	—
C^+	1.12 $\pm$ 0.05	13 $\pm$ 2
C^{2+}	2.08 $\pm$ 0.07	40 $\pm$ 3
C^{3+}	1.13 $\pm$ 0.05	73 $\pm$ 12
C^{4+}	0.10 $\pm$ 0.01	—
C^{5+}	< 0.01	—
ΣC^{n+}	4.43 $\pm$ 0.1	—
H^+	13.4 $\pm$ 0.3	31 $\pm$ 2
CH_3^+	< 0.1	—
CH^+	0.03 $\pm$ 0.02	—
CH_nI^+	< 0.1	—

[a] The peak of the recoil energy spectrum.

taking place in about 10^{-14} s, electrons are transferred from the neighboring atoms in the molecule to the highly charged iodine ion. As a result, the charges are distributed among all the atoms in the molecule and the molecule will subsequently explode due to the Coulombic repulsion. Although the model is simple, the recoil energy calculated from the net Coulombic repulsion reasonably agrees with experimental data (Table 2.3.2).

This model has later been checked by Shiokawa et al. [2.3.7], who measured the degraded ions as well as Br^{n+} resulting from the isomeric transition of ^{80m}Br in HBr, CH_3Br, and C_2H_5Br. An important observation from their experiments

Table 2.3.2. Comparison of experimental recoil energies with those calculated based on a Coulomb explosion model (Ref. 2.3.6)

Ion[a]	Experiment (eV)	Calculated (eV)
$I^{+4.8}$	8 $\pm$ 1	6
$C^{+2.05}$	41 $\pm$ 3	59
H^+	31 $\pm$ 2	50

[a] The charge in the average obtained from data in Table 2.3.1.

is that parent ions (HBr$^+$, CH$_3$Br$^+$, and C$_2$H$_5$Br$^+$, respectively) always constitute significant fractions of the fragmented ions. Furthermore, they have observed the existence of the ions with one H atom less than the parent ion. These findings may imply that the description of the model is essentially correct yet not in detail.

The molecular fragmentation has also been studied on β-decay of tritium incorporated in various compounds, such as HT and CH$_3$T [2.3.8]. The results predicted theoretically are in fairly good agreement with the experimental data and further suggest that the resulting ground state daughter ions HHe$^+$, LiHe$^+$ and BeHHe$^+$ are in bound state, while CH$_3$He$^+$, NH$_2$He$^+$, OHHe$^+$, and FHe$^+$ ions dissociate into a He atom and residual fragments [2.3.9].

For the studies on the mechanisms of energy loss during the slowing-down process and the identification of transient intermediates formed by reactions of the hot species with molecules of the medium, Wexler [2.3.1] employed two different types of mass spectrometers. In one method was studied the interaction of 0.8−3.75 MeV of protons with several isolated noble gas atoms and hydrocarbon molecules. On traversing of the protons through noble gases, ions in various charged states were formed: partial ionization cross sections were determined for the individual charge states. However, only singly-charged fragments were formed in collisions with polyatomic hydrocarbons. Another high pressure mass spectrometer is designed in search of transient ionic intermediates from the decay-induced exchange of tritium with CH$_4$. The mixture of T$_2$, D$_2$ and CH$_4$ of known composition was introduced into the source chamber (pressure in the chamber was about 0.1 mmHg). The results provided a good evidence for the ionic mechanism initiated by T^{3}He$^+$ ion, proposed by Pratt and Wolfgang [2.3.10].

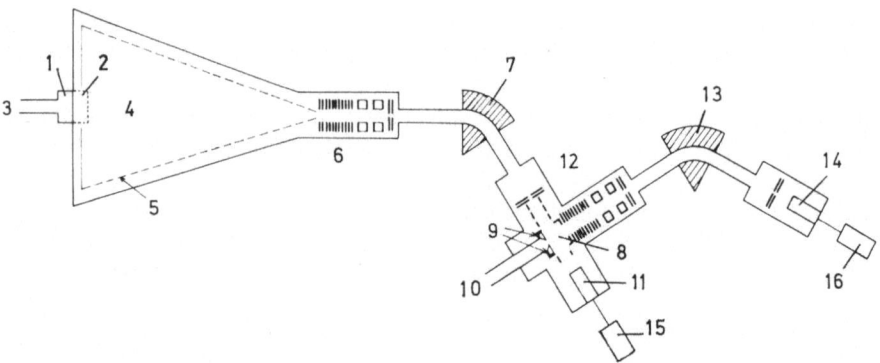

Fig. 2.3.3. Schematic diagram of TOHOKU charge spectrometer (Tohoku University) (Ref. 2.3.11).

1 X-ray tube or external ion source	*6* Lenses	*12* Lenses
2 Internal ion source	*7* Magnet	*13* Magnet
3 Gas inlet	*8* Reaction chamber	*14* Electron multiplier
4 Source volume	*9* Repeller	*15* Recorder
5 Guide rings	*10* Gas inlet	*16* Recorder or counter
	11 Electron multiplier	

A particular type of charge spectrometer has also been constructed, which consists of two mass spectrometers connected in perpendicular, as shown in Fig. 2.3.3 [2.3.11]. In its further modification, the amount of radioactive materials necessary for experiments is largely reduced by improving the efficiency of ion collection. This set-up serves not only as an ordinary charge spectrometer, but also as a double mass spectrometer for the study of ion-molecule reactions of decay-induced ions. The breakdown curves of alkyl ions has been determined, where the alkyl ions were produced by a charge exchange reactions with primary incident ions, such as $C_6H_6^+$, $C_2H_2^+$, Xe^+, He^+ [2.3.12].

2.3.2 Special Physical Techniques for Condensed Phase Studies

True chemical consequences of nuclear transformations in solid phase can hardly be elucidated by conventional wet-chemical means because of significant perturbations in the nature and distribution of the solid complexes [see Sects. 3.2 and 3.3]. Accordingly, direct physical techniques should be necessary for more complete understanding of the chemical states and surroundings of the recoil atoms immediately after their formation in solid phase. However, ordinary physical means such as ESR can never detect hot atoms whose concentrations are far below their detection limits. Hence, direct information of the hot atom reactions in solids can only be acquired by observing the chemical effects reflected on the radiations emitted from the recoil atoms. Mössbauer spectroscopy has been applied with considerable success to studies of hot atom reactions in solid state [2.3.13, 14]. Other physical techniques such as perturbed angular correlations, half life measurements, nuclear resonance fluorescence, etc. have so far provided only a limited amount of information regarding the chemistry of the recoil atoms in solids [2.3.14, 15].

Mössbauer Spectroscopy. Mössbauer spectroscopy consists in measurements of radiations (γ-rays, X-rays or electrons) emitted under recoilless conditions from nuclear levels of Mössbauer atoms resonantly excited by γ-rays with changing energies. Since the nuclear levels of the Mössbauer atoms are subjected to perturbation induced by their chemical states and surroundings, the Mössbauer spectra can provide detailed information as to their valence state, structure, site symmetry and magnetic properties. While in ordinary Mössbauer (absorption) measurements is examined the resonant absorption of γ-rays emitted by a standard radioactive source in unknown samples as absorbers, the absorption of γ-rays emitted by radioactive samples under study is measured against a standard absorber material in studies of solid phase hot atom chemistry. The latter technique is called Mössbauer emission spectroscopy, by which a hot atom chemist can acquire direct information as to what is going on about 10^{-7} to 10^{-8} s (i.e. order of the lifetimes of excited levels of Mössbauer atoms) after nuclear transformations producing hot atoms. For instance, ^{57}Co and ^{119m}Sn, two most commonly used Mössbauer source nuclides, decay via EC and IT to the Mössbauer levels, respectively, and chemical aftereffects of EC or IT decay may be observed on Mössbauer emission spectra of source compounds labeled (or doped) with ^{57}Co or ^{119m}Sn [see Sect. 3.3]. Besides with these radioactive nuclides, the Mössbauer level can also be populated via a nuclear reaction or Coulombic excitation. We may thus obtain further information regarding the displacement of hot source atoms in the lattice according to the recoil energies acquired in such reactions.

Ambe and coworkers have compared the behaviors of ^{119}Sn recoil atoms arising from the different nuclear processes in the lattice on the basis of ^{119}Sb- and ^{119m}Te Mössbauer emission spectra [2.3.16—18]:

$$^{119}\text{Sb} \xrightarrow{\text{EC}} {}^{119}\text{Sn}, \tag{2.3.1}$$

$$^{119m}\text{Te} \xrightarrow{\text{EC}} {}^{119}\text{Sb} \xrightarrow{\text{EC}} {}^{119}\text{Sn}, \tag{2.3.2}$$

$$^{120}\text{Sn}(\text{p}, 2\text{n})^{119}\text{Sb} \xrightarrow{\text{EC}} {}^{119}\text{Sn}. \tag{2.3.3}$$

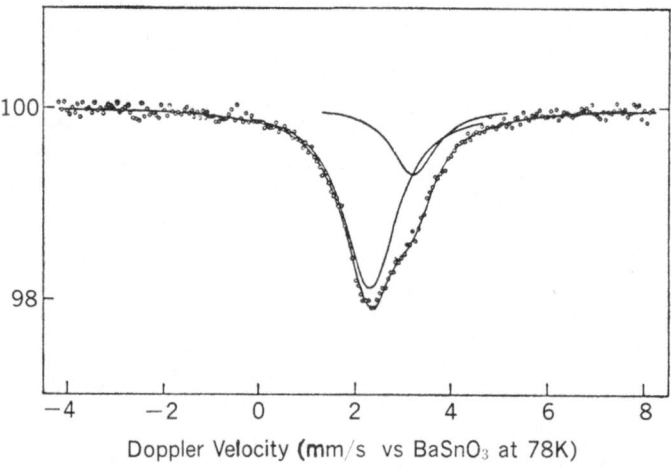

Doppler Velocity (mm/s vs BaSnO$_3$ at 78K)

Fig. 2.3.4. ^{119}Sn Mössbauer emission spectrum of Sb$_2^{119m}$Te$_3$ (78 K) (from Ref. 2.3.18). The minor peak (shoulder) has an isomer shift close to that in ^{119}Sn$_2$Te$_3$ and is attributed to recoil ^{119}Sn atoms stabilized on the Sb site; the major peak corresponds to ^{119}Sn retained on the Te site

Since two different sites exist in each of compounds such as Sb$_2$Te$_3$, SnSb, and SnTe, Mössbauer emission spectra of these compounds labeled with the source nuclide on a particular site will indicate whether or not the source nuclide has been eventually displaced by recoil from the original site in the lattice (Figs. 2.3.4 and 2.3.5). A comparison between the spectra of Sb$_2^{119m}$Te$_3$ (Fig. 2.3.4) and ^{119}Sb$_2$Te$_3$ reveals that a small fraction of ^{119}Sn atoms produced via double EC decay from ^{119m}Te have recoiled out of the original Te site and have been stabilized on the Sb site [2.3.18]. On proton-bombardment of ^{120}SnTe, ^{119}Sn atoms produced through process (2.3.) should acquire large recoil energies from (p, 2n) reaction. The Mössbauer emission spectrum of the proton-bombarded ^{120}SnTe demonstrates that about one-third of the ^{119}Sn atoms have been displaced from the original Sn site and occupy the Te site (Fig. 2.3.5) [2.3.17].

Time-differential Mössbauer emission spectroscopy, based on time-decayed coincidence measurements, is also a useful technique for the study of behavior of metastable species within a period of the order of lifetime of the Mössbauer level [2.3.19, 20]. For instance, Gütlich and coworkers have observed a metastable $^{57}Fe(II)$ species in anomalous spin state (S = 2) with a lifetime of about 100 ns at 77 K after EC decay in $[^{57}Co(phen)_3](ClO_4)_2 \cdot 2H_2O$ [2.3.21].

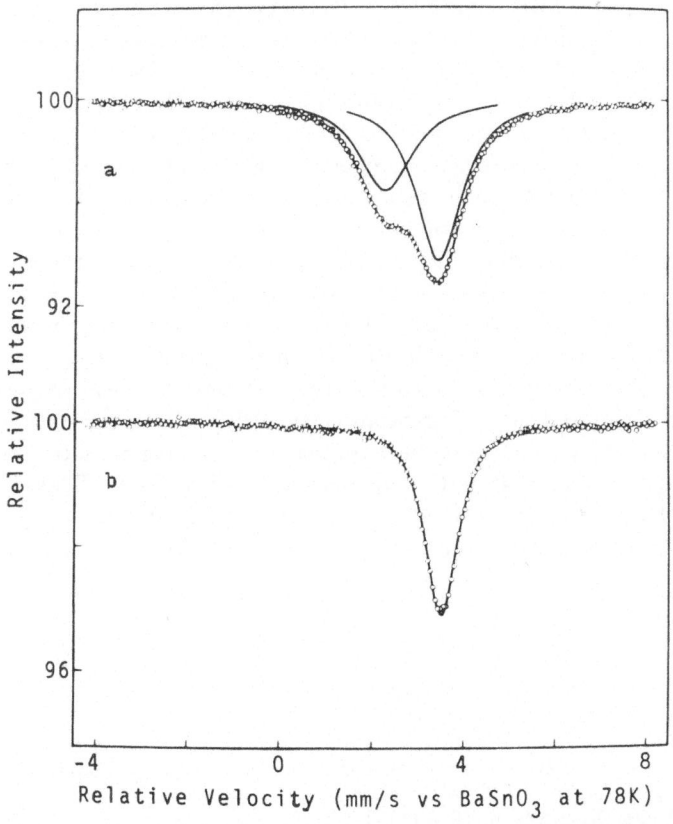

Fig. 2.3.5. ^{119}Sn Mössbauer emission spectrum (a) and ^{119}Sn Mössbauer absorption spectrum (b) of proton-bombarded $^{120}SnTe$ (from Ref. 2.3.17)

Perturbed Angular Correlation. In the absence of interactions with extranuclear fields, the angular correlation between two radiations (γ-rays) successively emitted by a single nucleus is given by the function

$$W(\theta) = 1 + \sum_{k\,even=2}^{k\,max} A_k P_k(\cos\theta) \tag{2.3.4}$$

where P_k are Legendre polynomials. If the extranuclear fields interact with the nucleus in the intermediate state between two emissions, however, the correlation is perturbed (due to the precession of the intermediate spin) and described by

$$W(\theta, t) = 1 + \sum_{\substack{k\,\text{even}=2}}^{k\,\text{max}} A_k G_k(t)\, P_k(\cos \theta) \tag{2.3.5}$$

where G_k is the time-dependent perturbation function.

Since the perturbed angular correlation (PAC) reflects the charge state and environment of the nuclei in the intermediate state, this technique can in principle provide hot atom chemists with information regarding the chemical effects produced by the nuclear transformations preceding the γ-transition in labeled sources. However, relatively few applications to hot atom studies have been attempted to date [2.3.22—24], probably because of considerable experimental difficulties in perturbed angular correlation measurements. For instance, measurements have been made of perturbed angular correlation of γ-cascades of ^{188}Os arising via β-decay from ^{188}Re in neutron-irradiated potassium rhenate KReO$_4$ [2.3.22]. While no hot atom effect can be detected by conventional wet chemical analysis of the irradiated solid KReO$_4$ after dissolution, the anisotropy in the perturbed angular correlation function (i. e., $\{W(180°)-W(90°)\}/W(90°))$ has changed appreciably after neutron irradiation and annealing of the solid sample, suggesting that a considerable fraction of the recoil atoms is displaced in the lattice. The information regarding the behavior of recoil atoms in solids can also be acquired by differential perturbed angular correlation experiments (DPAC) based on the measurements of perturbation (e.g. anisotropy) as a function of the time elapsed after the emission of the first γ-ray. Thus it has been suggested that the recoil ^{181}Hf atoms arising from neutron-irradiated hafnium complexes are eventually stabilized on two different sites in the lattice [2.3.24].

References

[2.3.1] Wexler, S.: Chem. Effects Nucl. Transform. IAEA, Vienna, Vol. 1, p. 3 (1965)
[2.3.2] Snell, A. H., Pleasonton, F.: J. Phys. Chem. *62*, 1377 (1958)
[2.3.3] Wexler, S.: J. Inorg. Nucl. Chem. *10*, 8 (1959)
[2.3.4] Wexler, S.: Chem. Effects Nucl. Transform. IAEA, Vienna, Vol. 1, p. 115 (1965)
[2.3.5] Carlson, T. A., White, R. M.: J. Chem. Phys. *38*, 2930 (1963)
[2.3.6] Carlson, T. A., White, R. M.: Chem. Effects Nucl. Transform. IAEA, Vienna, Vol. 1, p. 23 (1965)
[2.3.7] Takita, Y., Hiraga, M., Yoshihara, M., Shiokawa, T.: Radiochem. Radioanal. Lett. *7*, 313 (1971)
[2.3.8] Okuno, K., Sata, M., Yoshihara, K., Shiokawa, T.: ibid. *37*, 191 (1979)
[2.3.9] Ikuta, S., Yoshihara, K., Shiokawa, T.: J. Nucl. Sci., Technol. *14*, 720 (1977)
[2.3.10] Pratt, T. H., Wolfgang, R.: J. Am. Chem. Soc. *83*, 10 (1961)
[2.3.11] Shiokawa, T., Yoshihara, K., Yagi, M., Omori, T., Kaji, H., Hiraga, M., Nagatani, T., Takita, Y.: Mass. Spectr. *18*, 1230 (1970)
[2.3.12] Ikuta, S., Okuna, K., Yoshihara, K., Shiokawa, T.: Radiochim. Acta *25*, 21 (1978)
[2.3.13] Gütlich, P., Link, R., Trautwein, A.: *Mössbauer Spectroscopy and Transition Metal Chemistry*, Inorganic Chemistry Concepts Vol. 3. Berlin, Heidelberg, New York: Springer 1978

[2.3.14] Harbottle, G., Maddock, A. G. (eds.): *Chemical Effects of Nuclear Transformations in Inorganic Systems*. Amsterdam: North-Holland 1979

[2.3.15] Adloff, J. P.: Hot Atom Chemistry Status Report, IAEA-PL-615/5 (1975)

[2.3.16] Ambe, F., Ambe, S., Shoji, H., Saito, N.: J. Chem. Phys. *60*, 3773 (1974)

[2.3.17] Ambe, F., Ambe, S.: Chem. Phys. Lett. *39*, 294 (1976)

[2.3.18] Ambe, F., Ambe, S.: in *Isotopes and Chemistry* N. Saito (ed.). Tokyo: Hirokawa Publishing Co. 1978, p. 65

[2.3.19] Trifthäuser, W., Craig, P. P.: Phys. Rev. *162*, 274 (1967)

[2.3.20] Trifthäuser, W., Schroeer, D.: Phys. Rev. *187*, 491 (1969)

[2.3.21] Grimm, R., Gütlich, P., Kankeleit, E., Link, R.: J. Chem. Phys. *67*, 5491 (1977)

[2.3.22] Sato, J., Yokoyama, Y., Yamazaki, T.: Radiochim. Acta *5*, 115 (1966)

[2.3.23] Badica, T., Gelberg, A., Salageanu, S., Ion-Mihair, R., Ianovici, E., Zaitesva N. G.: Radiochim. Acta *16*, 36 (1971)

[2.3.24] Abbe, J. C., Marques-Netto, A.: J. Inorg. Nucl. Chem. *37*, 2239 (1975)

3 Characteristics of Hot Atom Reactions

3.1 Gas Phase Hot Atom Reactions

3.1.1 Theoretical Background

Kinetic Theory

When the system under study contains hot atoms with an initial energy E_0 in a thermal environment composed of one or more components, the hot atoms lose energy by collisions with all of the components. Below a maximum energy E_2 and above a threshold energy E_1, some fractions of the hot atoms may react on collision with one or more components to enter a chemical combination and thus be removed from the system.

The fundamental expression for the total probability P for a hot atom to react and enter a combination before thermalization is given by [3.1.1, 2]

$$P = \sum_j \int_{E_2}^{E_1} f_j p_j(E)\, n(E)\, dE \tag{3.1.1}$$

where f_j is the relative probability for collisions with component j, $n(E)\, dE$ the number of collisions between energy E and $E + dE$, and p_j the probability for reactive collision with component j. The essential problem is to find an appropriate expression for $n(E)\, dE$. Since such system is very similar to the neutron moderation in a fission reactor, Estrup and Wolfgang applied the neutron-slowing-down theory to Eq. (3.1.1) and obtained the following expression:

$$n(E)\, dE = (dE/\alpha E) \left[1 - \sum_j \int_{E_2}^{E} f_j p_j(E)\, n(E)\, dE \right] \tag{3.1.2}$$

where α is the average logarithmic energy loss per collision,

$$\alpha = - \left(\ln \frac{E(\text{after collision})}{E(\text{before collision})} \right). \tag{3.1.3}$$

It is further assumed that α varies rather slowly with the energy in the reaction region. Thus, in the case of a single reactant mixed with an inert moderator, the total yield of the hot reaction can be expressed by

$$P = 1 - \exp\left[(-f_j/\alpha)\, I \right], \qquad I = \int_{E_1}^{E_2} \frac{p(E)}{E}\, dE \tag{3.1.4}$$

where I is called "reactivity integral" and corresponds to the area under the excitation curve plotted on a logarithmic energy scale, and α is the weighed sum of the average energy loss on collision with reactant (α_{react}) and moderator (α_{mod}),

$$\alpha = f \cdot \alpha_{react} + (1 - f)\, \alpha_{mod}. \tag{3.1.5}$$

Combining Eqs. (3.1.4) and (3.1.5), Eq. (3.1.6)

$$-\frac{1}{\ln(1 - P)} = \frac{\alpha_{react}}{I} + \frac{\alpha_{mod}(1 - f)}{I f} \tag{3.1.6}$$

is derived. P is obtained experimentally at various moderator concentrations. On plotting $-[1/\ln(1 - P)]$ vs. $(1 - f)/f$, the slope gives α_{mod}/I and the intercept, α_{react}/I. The ratio of these two parameters corresponds to $\alpha_{mod}/\alpha_{react}$.

For the yield P_i of the ith individual hot product, Eq. (3.1.4) can be expanded as

$$P_i = \frac{f}{\alpha} I_i - \frac{f^2}{\alpha^2} K_i + \frac{f^3}{\alpha^3} L_i - \cdots \tag{3.1.7}$$

where

$$I_i = \int_{E_1}^{E_2} \frac{p_i(E)}{E}\, dE,$$

$$K_i = \int_{E_1}^{E_2} \frac{p_i(E)}{E} \left(\int_{E_1}^{E_2} \frac{p(E)}{E}\, dE \right) dE,$$

$$L_i = \int_{E_1}^{E_2} \frac{p_i(E)}{E} \left(\int_{E_1}^{E_2} \frac{p(E)}{E}\, dE \right)^2 dE.$$

Unless the total reactivity of the system is very large this series will converge rapidly. Normally, only the first two terms will be necessary, and in a system with very low reactivity the first alone may be sufficient. Comparing Eq. (3.1.7) with a series expansion of Eq. (3.1.4), the relations

$$\sum_i I_i = I, \quad \sum_i K_i = (1/2)\, I^2, \quad \sum_i L_i = (1/6)\, I^3, \ldots \tag{3.1.8}$$

are obtained. The terms K_i, L_i, are corrections for the first term I_i by considering "energy shadowing" which is the probability for the hot atom to have reacted above energy E forming a product. Hence, such terms will be small in a system with low reactivity. In a system with high reactivity, these terms are small for a product formed in the higher energy range but become large for products formed at lower energies.

The Estrup-Wolfgang kinetic treatments permit us to express the hot reactions in terms of important kinetic parameters, I and α. Furthermore, the comparison of the "energy shadowing" terms for the individual products provides at least qualitative information regarding the energy range where such products are formed. Some of the results obtained with the reaction of recoil tritium with simple hydrocarbons are listed in Table 3.1.1 [3.1.3].

Table 3.1.1. Relative values of kinetic parameters for reactions of recoil tritium[a] (from Ref. 3.1.3)

n-butane	$I = 3.34\ (\alpha_{He})$ $I_i\ (\alpha_{He})$	$\alpha = 4.5\ (\alpha_{He})$ $K_i\ (\alpha_{He}^2)$
HT	2.32	3.9
CH_3T	0.094	0.16
C_2H_3T	0.095	0.21
C_2H_5T	0.055	0.095
C_3H_5T	0.016	0.033
C_3H_7T	0.034	0.038
$n\text{-}C_4H_9T$	0.77	0.25
sum	3.38	4.69
$^1/_2 I^2$		5.28
n-pentane	$I = 4.38\ (\alpha_{He})$ $I_i\ (\alpha_{He})$	$\alpha = 5.9\ (\alpha_{He})$ $K_i\ (\alpha_{He}^2)$
HT	3.08	7.4
CH_3T	0.104	0.275
C_2H_3T	0.100	0.22
C_2H_5T	0.061	0.17
C_3H_5T	0.033	0.10
C_3H_7T	0.050	0.13
$n\text{-}C_4H_9T$	0.038	0.08
$n\text{-}C_5H_{11}T$	1.05	0.4
sum	4.52	8.71
$^1/_2 I^2$		9.59
neo-pentane	$I = 3.85\ (\alpha_{He})$ $I_i\ (\alpha_{He})$	$\alpha = 5.57\ (\alpha_{He})$ $K_i\ (\alpha_{He}^2)$
HT	2.40	6.0
CH_3T	0.295	0.55
C_2H_3T	0.030	0.10
C_3H_5T	0.015	0.0
$iso\text{-}C_4H_9T$	0.020	−0.05
pentenes-t	0.295	0.75
$neo\text{-}C_5H_{11}T$	0.85	−1.0
sum	3.91	6.35
$^1/_2 I^2$		7.41
ethane	$I = 2.30\ (\alpha_{He})$ $I_i\ (\alpha_{He})$	$\alpha = 3.04\ (\alpha_{He})$ $K_i\ (\alpha_{He}^2)$
HT	1.72	2.43
CH_3T	0.07	0.10
C_2H_3T	0.035	0.0
C_2H_5T	0.57	0.0
sum	2.40	2.63
$^1/_2 I^2$		2.65

Labeled compounds not listed in this table and 0.0 in the table indicates that such products were not detected in the analysis.

[a] I and α expressed in a unit of α_{He} which is the average logarithmic energy loss per collision with He atom.

A number of assumptions are actually involved in the development of the Estrup-Wolfgang theory in which recoil atoms lose energy and their energy distribution is altered by chemical reactions. It has been shown analytically and by computer simulation that the linear relation of the theory is also obtained even if all the assumptions necessary for the development of the theory are not fulfilled [3.4.4]. This indicates that the basic model used in the theory does not necessarily correspond to the physical reality. The kinetic theory is not reliable enough for estimating realistic parameters if the overlapping between probability functions is appreciable: the parameter which is most sensitive to overlapping is the K_i value for each product and the error in this parameter increases as α_{react} increases. It is also found in computer simulation that the energy distribution of hot atoms during the slowing down process has little effect on the derived kinetic parameters [3.1.5]. Under these conditions, the kinetic theory provides a means of estimating the parameters, α and I. The individual product kinetic parameters I_i and K_i, however, should be treated with caution since they generally include significantly greater uncertainties than the data from which they have been derived.

Despite such limitations of the theory, it is necessary to estimate the effect of the different energy loss parameters upon the yields of products. Amiel and coworkers have worked on these aspects [3.1.6, 7]. The relative value of α reflects the characteristics of the non-reactive collisions of hot atoms with partners. The interactions can be related with the scattering potentials. For an elastic transfer of energy, the relation between α and masses of the colliding particles is given by

$$\alpha/(1 - \beta) = \text{const.} \qquad (3.1.9)$$

if the scattering potentials $V(r) = A \cdot r^{-3}$ are the same for the different scatterers. In Eq. (3.1.9), $\beta = [(m - M)/(m + M)]^2$ and m and M are the masses of the hot atom and the target, respectively. Only the ratio of the α-values can be determined experimentally and is approximated by

$$\alpha_1/\alpha_2 = (1 - \beta_1)/(1 - \beta_2). \qquad (3.1.10)$$

Baer and Amiel [3.1.6] showed that the α_{m1}/α_{m2} values (subscripts 1 and 2 refer to different noble gases) were in a good agreement with the calculated values from Eq. (3.1.10) (Table 3.1.2) [3.1.6]. In collisions with polyatomic molecules, however, an upper limit for the α_r/α_m values (r and m refer to reactant and moderator) may be obtained by calculating α_r/α_m for all possible collisions with various parts of the molecule or with the molecule as a whole, and taking the largest ratio. In Table 3.1.3 are compared the calculated upper limits for α_r with the experimental results [3.1.7]. If the experimental α_r is smaller than the calculated maximum value in a system, all the energy losses of the hot atoms can be accounted for in terms of the elastic energy transfer, while otherwise there is a considerable contribution of the inelastic energy transfer. The latter effect is due to the symmetry of the target molecules: the interaction of a permanent dipole of such molecules and an induced dipole of the hot atom may cause the inelastic energy. However, it must be pointed out that the discrepancy between the experimental and the calculated values can be ascribed not only to the energy transfer in an inelastic collision, but also to the invalidity of Eq. (3.1.9) for elastic collisions in these cases.

Table 3.1.2. Ratios of energy loss parameters α_{m1}/α_{m2} for two different monoatomic moderators. (from Ref. 3.1.6)

System	$1-\beta$	α_{m1}/α_{m2}	
		Expl	Theoret
$^{80}Br-^4He$	0.180	0.14	0.180
$^{80}Br-^{82}Kr$	0.999		
$^{38}Cl-^4He$	0.345	0.41 ± 0.03	0.382
$^{38}Cl-^{20}Ne$	0.904		
$^3H-^{20}Ne$	0.452	0.42	0.461
$^3H-^4He$	0.980		
$^3H-^{40}Ar$	0.260	0.26	0.265
$^3H-^4He$	0.980		
$^{18}F-^4He$	0.595	0.59 ± 0.11	0.595
$^{18}F-^{20}Ne$	0.997		
$^{18}F-^{70}Ar$	0.874	0.94 ± 0.25	0.874
$^{18}F-^{20}Ne$	0.997		

Table 3.1.3. The average logarithmic energy losses of hot atoms (of mass m) colliding with reactants (of mass M) (from Ref. 3.1.7)

Reaction[a]	Experimental[b] α_r	Calculated upper limit of α_r	$M/(M+m)$
$^{18}F + CH_4(Ne)$	1.34 ± 0.36	1.0	0.47
$^{18}F + CF_4(Ne)$	1.47 ± 0.28	1.0	0.82
$^{39}Cl + CH_4(Ar)$	0.64^c	0.82	0.29
$^{39}Cl + C_2H_6(Ar)$	1.3^c	0.98	0.42
$^{80}Br + CH_4(Kr)$	0.25 ± 0.07	0.55	0.16
$^{38}Cl + CH_3Cl(Ar)$	2.3 ± 0.6	1.0	0.57
$^{38}Cl + CH_3Br(Ar)$	1.9 ± 0.4	0.87	0.71
$^{80}Br + CH_3Br(Kr)$	2.4 ± 0.6	1.0	0.54
$T + HD(Ar)$	3.8 ± 0.8	3.8	0.50
$T + CH_4(He)$	1.58 ± 0.30	0.78	0.84
$T + C_2H_6(Ne)$	5.69 ± 0.60	1.7	0.91
$T + C_3H_8(Ne)$	7.63 ± 0.77	1.7	0.93
$T + C_4H_{10}(Ne)$	8.00 ± 0.80	1.7	0.95
$T + C_4H_8(Ne)$	7.37 ± 0.74	1.7	0.95

[a] Moderator is in parentheses.
[b] In units of α (moderator).
[c] The value of the error was not given. The usual error in the experimental measurements of α is 15%—25%.

Dynamic Theory

Keizer [3.1.8] has developed a simple time-dependent theory of hot-atom reactions in a dilute host gas, on the basis of the Porter's stochastic theory [3.1.9] of hot-atom reactions. The stochastic theory denotes the average number of the atoms which have survived a single collision and have kinetic energy in the range dE', by

$$N_1(E, E')\, dE' = N \cdot P_1(E, E')\, dE' \tag{3.1.11}$$

where $P_1(E, E')\, dE'$ is the single collision survival probability for the energy range dE'. Assuming collision events to be independent and τ the mean free time, the energy distribution function for a hot atom having survived n collisions with surrounding bath gas can be characterized by the hot atom probability distribution $W_n(E, E')$ which described the probability of a recoil atom with the initial energy E and the energy E' after nth collision. With this definition, the average number of hot atoms lost between nth and (n + 1)th collisions is expressed by,

$$N_{n+1}(E) - N_n(E) = -N_n(E) \int W_n(E, E'')\, P(E'')\, dE'' \tag{3.1.12}$$

where $P(E'') = P_1(E'', E')\, dE'$ and $P_1(E'', E')\, dE'$, in turn, is the single collision survival probability for the energy range dE'. Changing to a continuous time scale by writing $n\tau = t$ and assuming that the mean free time is small, this equation becomes, after divided by τ,

$$d(N)/dt = -k(t)\, N(t) \tag{3.1.13}$$

where $k(t) = \tau^{-1} \int W(E, E't)\, P(E')\, dE'$, and the dependence on the initial energy has been made implicit.

Equation (3.1.13) would have the form of a chemical rate equation if the function k did not depend on time. This would occur, for example, if after a very short period of time the distribution function $W(E, E't)$ becomes time-independent — that is, has reached a steady state. Thus, when the steady-state probability distribution is rapidly approached, the loss of hot atoms is given by the first-order rate expression. This is so because if the reaction rate is much slower than the collision rate, then long before the number of hot atoms has been measurably depleted, the probability distribution function reaches a steady state, and Eq. (3.1.13) becomes

$$dN/dt = -kN \tag{3.1.14}$$

where $k = \tau^{-1} \int W(E, E')\, P(E')\, dE'$ and is time-independent. The k is the hot-atom rate constant which can be regarded as the average reaction probability per unit time at steady state. Since the reaction is bimolecular, k is proportional to the density of host molecules.

A more fundamental development of the above theory is based on the Boltzmann equation for the time variation of the hot atom probability distribution [3.1.10]. This provides a simple analytical method of describing and determining steady-state distributions in a hot-atom system.

The reservoir molecules, assumed as unperturbed by the hot atoms, are at thermal equilibrium at ambient temperature, T. The hot-atom reaction is written as

$$A + R \rightarrow AX + R',$$

in which A and R denote the hot-atom and reservoir species, and AX and R' the reaction products. Within the steady-state framework and assuming that the back reaction is negligible and the reaction does not significantly perturb the thermal properties of the reservoir, the following equation is given for the rate of change of the hot-atom density:

$$\frac{dN_A}{dt} = -N_A(t) \sum_j \int d\Omega \int dp_R \cdot \int dp_A \sigma^*(j) \, (p/\mu) \, g_A \cdot f_{R_j} e$$

$$= k(t) \, N_A(t) \tag{3.1.15}$$

where μ is the reduced mass of A and R, p the absolute value of their relative momentum, $\sigma^*(j)$ the cross section for the reaction channel, and g_A the probability density for the momentum p. The f_{R_j} is the distribution function for the reservoir molecules in the internal state j. Equation (3.1.15) has an appearance of the chemical rate expression, but it is complicated by the fact that the "rate constant" depends on the time implicitly through the time-dependence of the hot-atom momentum distribution, $g_A(p_A \cdot t)$. A reasonable first-order approximation for the shape of the distribution is Gaussian around the average value of zero momentum and this will necessarily have the form,

$$g_A(p_A \cdot t) = [2\pi m_A k T_A(t)^{-3/2}] \exp [-p_A^2/2m_A k T_A(t)]. \tag{3.1.16}$$

Then $(3/2) \, kT_A(t)$ is the average kinetic energy of an atom A. That is,

$$(3/2) \, kT_A(t) = \int dp_A \cdot (p_A^2/2m_A) \, g_A(p_A \cdot t). \tag{3.1.17}$$

This equation has the same form as the well-known "local equilibrium" distribution [3.1.11]. However, this differs from the standard treatment of local equilibrium in reactive systems in which the chemical reaction is assumed to provide only a perturbation on the bulk solvent, the Maxwell-Boltzmann distribution of the atoms may be quite different from the reservoir distribution, and this is accounted for by the use of hot-atom temperature in the local distribution.

The single physical parameter introduced by the local equilibrium distribution is the "temperature" of hot atoms, $T_A(t)$. In order to clarify the steady state momentum distribution it is necessary to find the steady states of the hot-atom temperature, which is attained by balancing the cooling and heating rates of atoms A. The cooling rate is determined by non-reactive collisions with the bath gas R, and heating occurs through the reactive collision which removes atoms A from the system at energies lower than $(3/2)kT_A(t)$.

In the steady state of hot-atom temperature T_A^*, where $dT_A^*/dt = 0$, there is spontaneous heating below T_A^* and spontaneous cooling above T_A^*.

The time derivatives of hot atom temperature follows from Eq. (3.1.17):

$$\frac{dT_A}{dt} = \frac{2}{3k} \left[\xi_n(t) + \xi_r(t) + \frac{3}{2} \, kT_A(t) \cdot \varkappa(t) \right]. \tag{3.1.18}$$

Equation (3.1.18), which specifies the coupling mechanism between reactive and non-reactive collisions, comprises the core of the local equilibrium modeling procedure. The $\xi_n(t)$, $\xi_r(t)$ and $\varkappa(t)$ follow directly from the Boltzmann equation, and are expressed by

$$\xi_n(t) = \sum_{j,k} \int d\Omega \int dp_A \, p_A^2/2m_A \int dp_R \sigma(k\,|\,j)\,(p/\mu)\times(g'_A f^{e'}_{Rk} - g_A f^e_{Rj}), \tag{3.1.19}$$

$$\xi_r(t) = \sum_j \int d\Omega \int dp_A p_A^2/2m_A \int dp_R \sigma^*(j)\,(p/\mu)\,g_A \cdot f^e_{Rj}, \tag{3.1.20}$$

$$\varkappa(t) = \sum_j \int d\Omega \int dp_A \int dp_R \sigma^*(j)\,(p/\mu)\,f^e_{Rj} g_A. \tag{3.1.21}$$

In Eqs. (3.1.19—21), j and k specify the initial and final quantum states of the reservoir molecules. The momentum density distribution, f^e_{Rj}, for the reservoir molecules is Maxwellian at ambient temperature (T). The $\sigma^*(j)$ and $\sigma(k/j)$ denote reactive and nonreactive collision cross sections.

Assuming that, at the limit of large t, $T_A(t)$ approaches a steady state value T_A^*,

$$\lim_{t\to\infty} \frac{dT_A}{dt} = \frac{dT_A^*}{dt} = 0 = \frac{2\varkappa^*}{2k}\left(\frac{\xi_n^*}{\varkappa^*} + \frac{\xi_r^*}{\varkappa^*} + \frac{3}{2}\,kT_A^*\right) \tag{3.1.22}$$

where ξ_n^*, ξ_r^* and $\varkappa^*$ are given by Eqs. (3.1.19—21) with g_A taken equal to g_A^*, the steady state distribution. Thus, Eq. (3.1.22) gives T_A^* implicitly in terms of the quantities ξ_n^*, etc. which depend on T_A^* through g_A^*.

As for the physical meaning of the various terms in Eq. (3.1.22), $\varkappa(t)$ is the time-dependent quasi-rate constant, k^* indicates its steady-state value, and the rate equation can be written as

$$dN_A/dt = -\varkappa^* \cdot N_A(t). \tag{3.1.23}$$

Since ξ_n gives the rate of change per atom of the hot atom kinetic energy due to non-reactive collisions, ξ_n^* denotes the rate at steady state. The ratio $\xi_r^*/\varkappa^*$ is the reactive collision term, and can be expressed as $\langle\varepsilon\rangle_r$, the average energy of a reacting hot atom from Eqs. (3.1.20) and (3.1.21). Thus, the steady-state temperature can now be written as

$$-\frac{3}{2}\,kT_A^* + \langle\varepsilon\rangle_r = \xi^*/\varkappa^*. \tag{3.1.24}$$

T_A^* of Eq. (3.1.24) can be obtained with the required knowledge of the cross sections for various collision processes occurring in the systems.

Root et al. [3.1.12] have applied this steady-state treatment to multicomponent reaction systems, i.e. Ar-moderated $^{18}F-H_2$ systems. In the calculations, Eq. (3.1.19) has been approximated by combining derived expressions for the hard-sphere elastic energy loss and for the collision frequency between hot atoms and Maxwellian equilibrium reservoir molecules. The theoretical reactive cross sections obtained for quasi-classical trajectory calculations have also been used.

Equation (3.1.25), inverted formula of Eq. (3.1.18), was integrated numerically over T_A intervals ranging from 10^7 K to ambient temperature:

$$\Delta t = \int_{T_A^i}^{T_A^f} F^{-1}(T_A) \cdot dT_A \tag{3.1.25}$$

where $F(T_A)$ denotes the right-hand side of Eq. (3.1.18), while indices i and j specify the initial and final conditions. Since $N_A(t)$ can be obtained by integration of Eq. (3.1.26),

$$\varkappa(t) = -\left(\frac{t}{N_A(t)}\right)\left(\frac{dN_A(t)}{dt}\right), \tag{3.1.26}$$

the reaction yield can be calculated according to Eq. (3.1.27):

$$Y_{(t^f)} = \int_0^{t^f} \varkappa(t)\, N_A(t)\, dt. \tag{3.1.27}$$

The reaction yield obtained has been shown in Table 3.1.4 together with the mean reactive life-time, which is given by Eq. (3.1.28). The results exhibit a pronunced negative dependence upon Ar-moderation, characteristic of hot atom moderator experiments.

$$\tau = \frac{\int_0^{t^f} t\varkappa(t)\, N_A(t)\, dt}{\int_0^{t^f} \varkappa(t)\, N_A(t)\, dt}. \tag{3.1.28}$$

Table 3.1.4. Total hot yield and mean reactive lifetime results for the Ar moderated non-thermal $^{18}F + H_2$ reaction systems (taken from Ref. 3.1.12)

Argon concentration (mole %)	Total hot yield (300 K) (%)	Total hot yield (10 K) (%)	τ^a (molecule s cm^{-3})	$\langle t^2 \rangle^a$ $\times 10^{-9}$
0.0	99.5	99.0	1.64	0.51
10.0	98.5	97.4	1.70	0.60
30.0	93.3	90.3	1.86	0.76
50.0	81.1	75.8	1.99	0.86
70.0	58.6	52.4	2.07	0.89
90.0	23.2	19.8	2.10	0.88
99.0	2.5	2.1	2.10	0.87

[a] Calculated for 10 K ambient temperature.

3.1.2 Gas Phase Hot Atom Reactions

When the energy of the reacting atoms and molecules obeys the Maxwell-Boltzmann distributions, the rate constant for such process is expressed in formula of the form,

$$k = A \exp(-E/RT). \tag{3.1.29}$$

Investigations of chemical kinetics using reaction species with Maxwellian distributions offer information concerning the behavior averaged over a distribution strongly weighted on lower energy side but give little insight into factors which are important at higher energies. Figure 3.1.1 shows hypothetical reaction cross sections for possible reactions of atomic hydrogen with a simple hydrocarbon, RH. The study of energetic atoms formed in nuclear reactions has permitted rapid progress in outlining the nature of such energetic reactions. In the following sections, the characteristics of thermal and energetic atom reactions will be discussed.

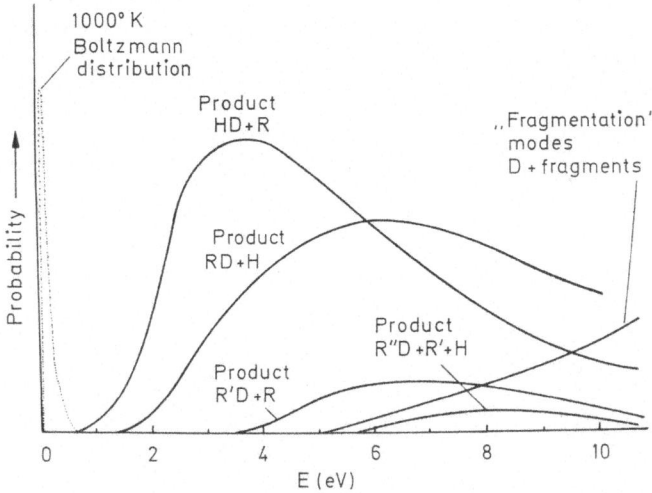

Fig. 3.1.1. Hypothetical energy dependences in D + RH system (taken from Ref. 3.1.1)

Thermal Atom Reactions

The velocity distribution of atoms in thermal equilibrium at a given temperature is expressed by the Maxwellian distribution,

$$\frac{dn}{N_0} = 4\pi \left(\frac{m}{2\pi kT}\right)^{3/2} v^2 \cdot \exp\left(-\frac{mv^2}{2kT}\right) dv. \tag{3.1.30}$$

Figure 3.1.2 shows the velocity distribution of hydrogen atoms at 300 K. Numbers of reactive atoms at particular energies are also included.

The characteristics of thermal reactions can be understood from the viewpoint of the reaction probability distribution. The total yield of a particular product in that reaction system can be expressed simply by

$$P(v) = \int_0^\infty n(v)\,\sigma(v)\,dv \qquad (3.1.31)$$

where $\sigma(v)$ is the reaction cross section for an atom with velocity v, and $n(v)\,dv$ indicates the number of molecules having velocity between v and $v + dv$. Since the knowledge regarding reaction cross section is extremely primitive, it seems convenient to assume σ simply as follows:

$$\sigma(v) = k(v - v_0), \qquad v \geqq v_0 \qquad (3.1.32)$$

$$\sigma(v) = 0, \cdot \qquad\qquad v < v_0 \qquad (3.1.33)$$

in which v_0 is the velocity at the threshold energy for a particular reaction. Figure 3.1.2 also includes the calculated reaction probability at $v_0 = 0.8 \times 10^6\,\mathrm{cm\,s^{-1}}$ and $1.39 \times 10^6\,\mathrm{cm\,s^{-1}}$, corresponding to about 0.24 and 1.0 eV for hydrogen atoms. These curves show that the probability of reaction for a Maxwellian distribution rapidly reaches a maximum not too far above the threshold energy. Another important observation is that the total reaction probability falls off very quickly with an increase in the threshold energy.

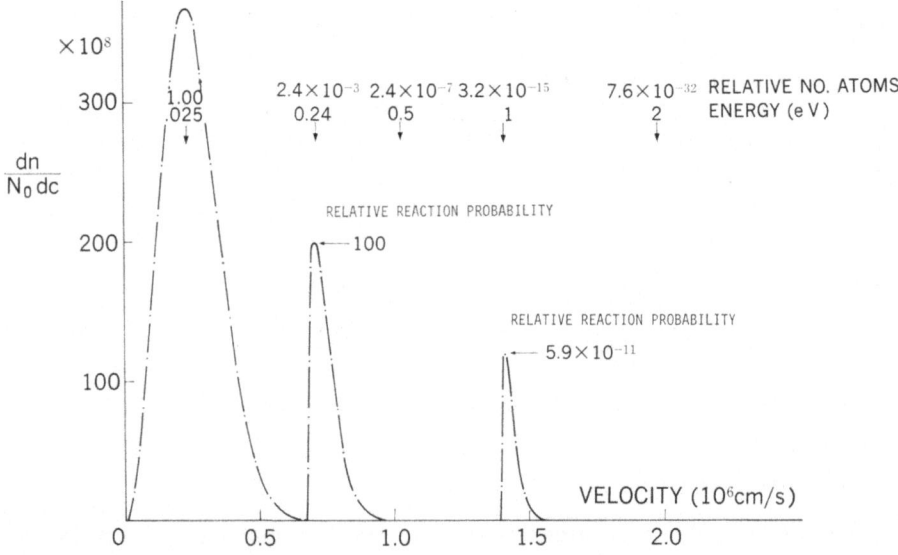

Fig. 3.1.2. Maxwellian distribution of hydrogen atoms at 300 K

Energetic Atom Reactions

In most thermal reactions, the actual rates and modes of reactions observed are strongly dependent upon the activation energies. On the other hand, the recoil atoms of initial energy E_0 lose energy by collisions with all of the components in the system, and approach this reaction range from the high energy side. Consequently there are non-zero probability for various kinds of reactions. The reaction probability P for the recoil atoms reacting to enter chemical combinations in the energy range $E_2 - E_1$ can then be expressed by Eq. (3.1.1).

The study of the reactions of energetic atoms is of interest for two reasons: one is to identify new reactions which are not observed with thermal atoms, and the other is to obtain quantitative information on the reaction cross section as a function of energy, and on the reaction threshold for novel reactions.

Characteristics of Energetic Reactions

In an energetic reaction system, the observable hot interactions should always have a resonably high reaction cross section in order to give a measurable amount of product. Thus, a hot interaction with the bulk system should not be affected by the presence of a small amount of an additive even if the reaction of the hot atoms with the additive is extremely efficient. On the other hand, observable thermal reactions may have a low reaction probability towards ordinary substances and can be completely suppressed by the presence of a minor additive with a high reaction efficiency. Such an efficient additive is called scavenger.

A moderator is also used sometimes to change the energy spectrum of the energetic atoms, or to lower the average energy of the reacting atoms. Inert gases such as He, Ne, Ar, Kr, and Xe are generally employed as moderators. While they remove translational energy of hot species through elastic collisions, they do not affect the thermal energy distribution of the system.

In general, primary interactions of energetic atoms should be insensitive to both temperature and pressure. However, if sufficient energy is left in the primary reaction products to undergo secondary changes, the dependence of the product yield on pressure and temperature will be observed [3.1.13].

Although the reactions of mono-valent hot atoms vary from element to element, the reactions display very similar basic features. In the following are described the basic reaction modes of recoil tritium, which are applicable to most hot atoms with certain modifications [3.1.14—18].

a) Abstraction of an atom:

$$T^* + RX \rightarrow TX + R \quad \text{(shown as T-to-TX reaction)}.$$

This is normally the most efficient reaction mode of hot atoms. However, in many cases, especially when X is a halogen atom, the quantitative measurement of the reaction yield will be difficult. The HT formed by T-to-HT reaction is almost the only observable product.

b) Substitution of an atom:

$$T^* + RX \rightarrow RT + X \quad \text{(shown as T-for-X reaction)}.$$

This reaction usually has a much higher threshold energy than the abstraction reaction. With hot atoms, the reaction is sometimes as important as the abstraction.

c) $T^* + RX \rightarrow R_1T + R_2X \quad \text{(shown as T-for-alkyl reaction)}.$

This reaction generally has much lower efficiency than the substitution of an atom. As a general trend, the hot atom will combine with the smaller radical formed upon rupturing of the C—C bond in the reactant molecule.

"Double substitution reaction" has been proposed as a possible reaction path for recoil tritium with organic molecules; e.g.,

$$T^* + RH \rightarrow R_1T + R_2 + R_3.$$

However it is usually concluded that this reaction is only a minor reaction mode for most hot atoms.

d) Addition to an unsaturated bond:

$$T^* + CH_2{=}CH_2 \rightarrow [CH_2T{-}CH_2]^*.$$

The reaction itself is usually an exothermic and the resulting radical may be left with an excitation energy enough to undergo unimolecular decomposition, e.g.,

$$[CH_2T{-}CH_2]^* \rightarrow CHT{=}CH_2 + H.$$

Unimolecular Decomposition

In the substitution reactions of hot atoms, most of the excess energy originally possessed by the atom will be transferred into the products in a vibrational energy mode. Thus, the resulting molecule can either be stabilized by losing such excess energy through collisions with surrounding molecules, or decompose unimolecularly. The extent of collisional stabilization is proportional to pressure. Thus, the competition of two subsequent reaction modes of the primary substitution products can usually be examined by determining the yields of two reaction products as a function of pressure.

With a further assumption that the internal energy had been randomly distributed in the whole molecule before decomposition [3.1.19], the rate of decomposition can be correlated with the excitation energy by applying the RRKM (Rice-Ramsberger-Kassel-Marcus) theory [3.1.20].

Stereospecificity of the Substitution Reactions

When an atom or a radical is substituted for by a hot atom, there exist two possible approaches: a forward attack and a backward attack. It may be expected that the configuration of the substrate molecule is retained in the former type of reaction, while it is inverted in the latter. Hence, information can be obtained experimentally through the use of optically active molecules containing asymmetrical carbons. It is usually considered that substitution reactions are accompanied by essentially complete retention of the original configuration. Recently, however, an evidence has been reported for the substitution reaction of recoil chlorine atoms with inversion in both the gaseous and condensed phases [3.1.21].

Isotope Effects

Although the kind of isotope effect observable in ordinary thermal reactions still has some importance in hot reactions, different types of isotope effects become predominant in the latter.

Since the hot reaction yield is equal to the integration of the product of the reaction probability per collision, p(E), and the collision density, n(E), over the entire reaction range as expressed by Eq. (3.1.1), the isotope effect may result from a difference in any of p(E), n(E), and integration range $(E_1 - E_2)$, for isotopically labeled molecules. They are called as "reactive isotope effect", "moderator isotope effect", and "average energy isotope effect", respectively (Fig. 3.1.3) [3.1.22].

a) Moderator isotope effect: It is possible that the average energy loss in non-bonding collisions is different for two isotopic molecules. The situation is illustrated in Fig. 3.1.3a in which the average energy loss in collision with A is less than that with B. This is equivalent to saying that the number of collisions available in the reaction energy range is larger in A than in B. A higher hot yield is expected from the reaction with A than with B, if the reaction probabilities per collision are comparable.

b) Reactive isotope effect: In the absence of the moderator isotope effect, the isotopic variation is due to the difference in the reaction probabilities per collision at a given energy between the two isotopically labeled molecules [Fig. 3.1.3b, c].

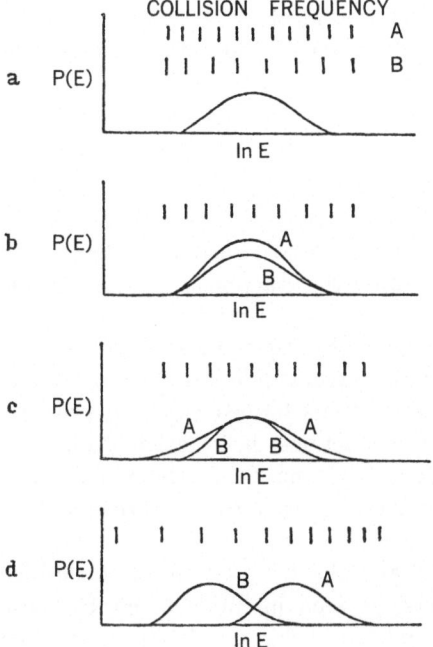

Fig. 3.1.3 a—d. Possible sources of isotopic variation in recoil tritium experiments. a Moderator isotope effect, $\alpha_A < \alpha_B$. b, c Probability integral isotope effects — greater reaction probability per collision for A vs. B (a); greater range for reaction for A vs. B (b). d Average energy isotope effect, reaction with B occurs at lower energies, for which the tritium flux has been reduced by reaction at higher energies. (Reprinted with permission from Ref. 3.1.22. Copyright (1965) American Chemical Society)

c) Average energy isotope effect: This type of isotope effect is illustrated in Fig. 3.1.3d, in which the flux of hot atoms available at lower energies has been appreciably diminished by higher energy reactions. Fewer hot reactions will be observed with B than with A, even if the areas under the reaction probability curves are of equal magnitude.

In order to make a more reliable measurement of the reactive isotope effect, it is always desirable to reduce the moderator- and the average energy-isotope effects to a much lesser extent. Such requirement is usually attained by selecting proper experimental conditions: (i) choose as reactant(s) a partially labeled molecule such as CH_2D_2, or a mixture of two isotopic molecules such as CH_4/CD_4, or (ii) employ a third molecule as a bulk component in the reaction system.

In case (i), the observed isotope effect is completely free from the moderator isotope effect. In case (ii), the energy spectrum of the reacting atoms is predominantly controlled by the non-reactive collisions with the bulk component and not significantly affected by the choice of the isotopic molecules. Furthermore, especially when the inert gas was chosen as a third molecule, total hot reaction yield is largely reduced but the flux is not appreciably diminished in the higher energy range. The average energy isotope effect is essentially absent.

The reactive isotope effect can be further classified into the primary and secondary isotope effects according to the nature of the reacting bonds and the composition of the reacting molecules on which the abstraction or substitution takes place. The primary isotope effect originates from the difference in the identity of the atoms being abstracted or replaced. On the other hand the isotopic atoms in non-reacting bonds may also affect on the reactivity of the atom in the same molecule. This is called the secondry isotope effect. Within the limited information obtained, both primary and secondary isotope effects favor the protonated over the deuterated compounds.

3.1.3 Tritium

Recoil tritium is most conveniently generated by slow neutron bombardment of ^6Li or ^3He. Several reviews have appeared on the reactions of recoil tritium [3.1.1, 2, 14—18].

In Table 3.1.5 are listed the scavengers commonly employed for recoil tritium reactions. Among these, oxygen may be the best to avoid undesirable complication. However, if one needs to trap tritiated free radicals formed either in primary hot atom interactions or in some secondary reactions, it is rather preferable to adopt alternative scavengers such as H_2S, Br_2, and I_2. In these cases, the tritiated radicals are nearly quantitatively transformed into the corresponding hydrocarbons, bromides, and iodides.

In the reactions with hydrocarbons, HT and the tritiated parent molecule, RT, are always two predominant products, as seen in Table 3.1.6a, b. With unsaturated compounds, the tritiated parent molecules are formed not only

Table 3.1.5. Commonly employed scavengers for recoil tritium reactions (taken from Ref. 3.1.18)

Scavenger	Phases employed in	Can scavenged products be quantitively measured	Additional remarks
O_2	Gas	No	Very commonly used for gaseous samples
NO	Gas	No	Scavenged product may decompose to give additional olefin yields
Br_2	Gas, Liquid	Yes (Bromides)	Add to double and triple bonds
	Gas, Liquid	Yes (Iodides)	1) Add to double and triple bonds 2) Low concentration in the gas phase 3) Some Iodides quench the counter
HI	Gas, Liquid	Yes (Hydrocarbons or HT)	Normally freshly prepared
H_2S	Gas, Liquid	Yes (Hydrocarbons or HT)	Introduce some ^{35}S radioactivity
SO_2	Gas, Liquid	No	1) Introduce some ^{35}S radioactivity 2) Relatively, inefficient
ICl	Gas, Liquid	Yes (Halides)	1) Add to double and triple bonds 2) Decompose to yield I_2 and Cl_2 readily
Olefins	Gas, Liquid	Partially (Olefins)	Relatively, inefficient
DPPH	Liquid	No	Inefficient

through T-for-H reaction (Eq. (3.1.34)) but also through the addition-decomposition processes as shown in Eq. (3.1.35):

$$T^* + CH_2{=}CH_2 \rightarrow CH_2{=}CHT + H, \tag{3.1.34}$$

$$T^* + CH_2{=}CH_2 \rightarrow [CH_2T{-}CH_2]^*,$$

$$[CH_2T{-}CH_2]^* \rightarrow CHT{=}CH_2 + H. \tag{3.1.35}$$

On the per bond basis, addition reaction is much more efficient than the direct substitution of vinyl hydrogen. The HT/RT ratio is usually much less than that obtained with alkanes with similar carbon skelton.

Table 3.1.6a. Relative yields of products from recoil tritium reactions with alkanes in well-scavenged systems (from Ref. 3.1.18)

T-labeled products	Target molecules							
	CH_4	C_2H_6	C_3H_8	C_4H_{10}	i-C_4H_{10}	C_5H_{12}	i-C_5H_{12}	neo-C_5H_{12}
HT	79	165	185	197	191	237	220	128
CH_3T	100	11.5	10.9	7.5	15.8	7.2	14.2	20.1
C_2H_5T	Tr	100	⎫ 12.3	3.9	0	4.4	2.8	0
C_2H_3T	—	—	⎭	5.9	1.0	7.2	3.0	0.3
C_3H_7T	—	—	100	3.1	3.2	4.6	1.4	0
C_3H_5T	—	—	—	1.8	7.0	1.8	4.8	0.9
$C_4H_{11}T$	—	—	—	100	0	2.7	1.7	0
i-C_4H_7T	—	—	—	0.5	0	Tr	1.2	0
i-C_4H_9T	—	—	—	—	100	0	1.0	1.2
i-C_4H_7T	—	—	—	—	1.0	0	Tr	4.2
$C_5H_{11}T$	—	—	—	—	—	100	0	0
i-$C_5H_{11}T$	—	—	—	—	—	—	100	0
neo-$C_5H_{11}T$	—	—	—	—	—	—	—	100
CH_2T	20 ⎫	⎫ 44.2	⎫	—	—	—	—	8.2
C_2H_4T	— ⎭	⎭	7.2	—	—	—	—	—
C_3H_6T	—	— ⎭	⎭	—	—	—	—	—

In halogenated compounds, T-for-X (X = I, Br, Cl, F) reactions also occur and compete with T-for-H reactions with less efficiency. Two general trends are observed: One is that the relative efficiency of T-for-X reaction decreases in the order: I > Br > Cl > F. Another important feature is the extensive decomposition of primary halogenated products, because of the relative easiness of the HX elimination.

Recoil tritium also reacts with silane and other organosilicon compounds: the reaction processes follow the general characteristics of recoil tritium, and both abstraction and substitution occur at Si—H bond [3.1.23], even with higher efficiency. However, the total reactivities [3.1.24] of $(CH_3)_4C$ and $(CH_3)_4Si$ appear to be approximately equal, with the former parhaps 10% larger. This reveals that the latter behaves very similarly to hydrocarbons, with respect to reactivity as well as in sensitivity toward oxygen. The total reactivity of $(CH_3)_3SiH$ toward abstraction is not quite different ($\pm 15\%$) from that of $(CH_3)_3CH$, suggesting that abstraction from tertiary C—H bond is quite similar to that from tertiary Si—H bond.

Pressure Dependence of Substitution Product Yield (Median Excitation Energy)

In unimolecular decomposition of excited primary products formed by recoil tritium substitution, the most dominant mode is almost always identical to the one prevalent in thermal systems. A consecutive decomposition sequence is also observed with certain molecules [3.1.25], such as $[C_2H_3TF_2]^*$,

$$[C_2H_3TF_2]^* \rightarrow [C_2H_2TF]^* + HF, \tag{3.1.36}$$

$$T^* + CH_3CHF_2 \rightarrow [C_2H_3TF_2]^* + H, \tag{3.1.37}$$

Table 3.1.6b. Relative yields of products from recoil tritium reactions with alkenes in gaseous systems (Ref. 3.1.18)

T-labeled products	Target molecules					
	C_2H_4	C_3H_6	1-butene	trans-2-butene	cis-2-butene	iso-butene
HT	23	72	141	176.7	167.8	88.3
CH$_3$T	0	—	5.9		8.2	4.2
C$_2$H$_5$T	0	—	2.1	3.3	0.2	0
C$_2$H$_3$T	100	36	41.2		3.7	0
C$_2$HT	3.5	—	2.3	—	2.0	0.7
C$_3$H$_7$T	0	—	0	—	0	0
C$_3$H$_5$T	0	100	46.7	16.2	66.4	1.9
Allene-t	0	0	1.0	0	0.7	2.5
C$_4$H$_9$T	0	—	0	26	8.7	0
1-butene-t	—	—	100	18.7	10	—
trans-2-butene-t	—	—	4.2	100	10.1	—
cis-2-butene-t	—	—	3.7	5.8	100	—
iso-butene-t	—	—	—	—	—	100

$$[C_2H_2TF]^* \rightarrow CH{=}CT + HF. \tag{3.1.38}$$

Systematic studies have been made on unimolecular decomposition of the substitution products from cyclobutane [3.1.26]. The reaction sequence for the T-for-H reaction is shown below:

$$T^* + c\text{-}C_4H_8 \rightarrow [c\text{-}C_4H_7T]^* + H, \tag{3.1.39}$$

$$[c\text{-}C_4H_7T]^* + M \xrightarrow{(S)} c\text{-}C_4H_7T + M, \tag{3.1.40}$$

$$[c\text{-}C_4H_7T]^* \xrightarrow{(D)} C_2H_3T + C_2H_4. \tag{3.1.41}$$

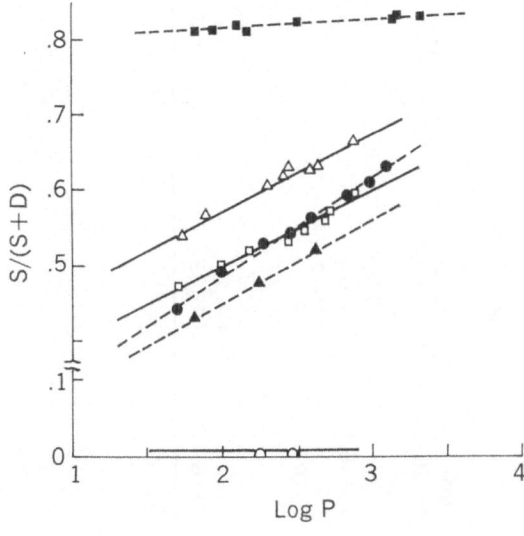

Fig. 3.1.4. Pressure dependence of S/(S + D) for the T-for-H substitution in various systems (Ref. 3.1.18)

■ CH$_4$
□ c-C$_4$H$_8$
△ i-C$_3$H$_7$Cl
○ CH$_3$NC
● C$_2$H$_5$Cl
▲ c-C$_4$D$_8$

Table 3.1.7. Estimates of median energy deposition for energetic atom reactions (taken from Ref. 3.1.17)

Reaction	Substrate	Primary product	Secondary product	Act. energy (eV)	Estimated median energy (eV)
T/H	c-C₄H₈	c-C₄H₇T	C₂H₃T	2.7	5 ± 1
	CH₄	CH₃T	CH₂T	4.5	3.5 ± 1
	CH₃NC	CH₂TNC	CH₂TCN	1.6	> 4
	CH₃CF₃	CH₂TCF₃	CHT=CF₂	3.0	4.7 ± 1
	spiro-pentane	⟨spiro-C₅ structure: CHT with two CH₂⟩	CHT=C=CH₂ or CHT=CH₂	2.5	? 5
T/D, T/CH₃	c-C₄D₈ ⟨—CH₂—CH₂ ring⟩	c-C₄D₇T ⟨CH₂ ring structure⟩	C₂D₃T	2.7	5 ± 1
T + olefin	CH₃CHCH₂CHCH₃	CH₃CHCH₂CHT	CH₂=CHT	? 2.7	6 – 8
	1-butene	CH₃CH₂CHCH₂T	CH₂TCH=CH₂	1.4	0.5
	cis-2-butene	CH₃CHTCHCH₃	CH₃CH=CHT	1.4	0.5
¹⁸F/F	CF₄	CF₃¹⁸F	CF¹⁸F	9.6	> 10
	CH₃CF₃	CH₃CF₂¹⁸F	CH₂=CH¹⁸F	3.0	8.3 ± 1.5
	c-C₄F₈	c-C₄F₇¹⁸F	several	3.2	> 10
	c-C₃F₆	c-C₃F₅¹⁸F	CF₂=CF¹⁸F	1.7	≥ 10 ± 2
¹⁸F/H	CH₃CF₃	CH₂¹⁸FCF₃	CH¹⁸F=CF₂	? 3.0	6.3 ± 1.0

The relative yield of stabilization (S) increases with the increasing pressure, while that of decomposition (D) decreases in the meantime. In Fig. 3.1.4, the fraction stabilized, (S)/(S + D), is plotted versus the logarithm of the effective pressure, together with those obtained in other reaction systems. With the assumption of internal energy randomization in the molecule before its decomposition, the RRKM theory was applied to the decomposition of $[c\text{-}C_4H_7T]^*$, the results show that the T-for-H reaction in $c\text{-}C_4H_8$ leaves the products with a broad distribution of internal energy with a median at about 5 eV [3.1.26].

A similar treatment has been extended to various reaction systems, and the results obtained are summarized in Table 3.1.7 [3.1.17]. Mechanistic considerations indicate that the reactive collision is cushioning and some transient state is formed in the course of reactions, and that the state is very different from the ordinary transient state and the deposited energy is a good measure of the excess energy of the hot atom [3.1.14].

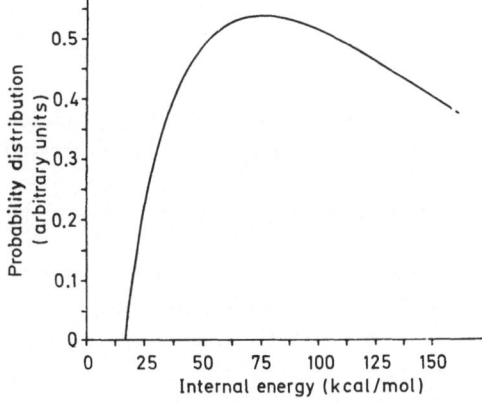

Fig. 3.1.5. Relative distribution of internal excitation energy in cyclo-butane-T produced by kinetically controlled chemical activation in the nuclear recoil tritium for hydrogen-substitution reaction (taken from Ref. 3.1.28)

The excitation energy of the radicals formed by addition of recoil tritium to olefins is equal to the sum of the heat of reaction and the excess energy of the tritium at the time of reaction. The total excitation energy indicates that most addition reactions occur at tritium kinetic energy less than 0.5 eV [3.1.27].

As an attempt to estimate a complete energy distribution of excited molecules prior to unimolecular decomposition, a hypothetical energy spectrum was proposed to fit the experimental data [3.1.14]. A calculation based on a Monte Carlo optimization [3.1.28] technique predicts a similar characteristic feature with the same sets of data and assumptions (Fig. 3.1.5).

Stereochemistry of T-for-H Reactions

Since the studies on the retention/inversion in the solid phase, experiments with various substrates have been carried out in the gas phase [3.1.29]. All these results indicate almost complete retention of configuration in the T-for-H reactions. However, it must be considered that in the experiments, the molecules must contain asymmetric carbon atom carrying three heavy substituents. Thus, it is alway possible that the backside attack by a hot atom is sterically hindered and only front-attack by a hot atom is permitted to occur. In order to explain this question, molecules with several light substituents, particularly CH_4, are used as the most favorable substrates. The trajectory calculation with CH_4 shows that substitution with inversion is indeed feasible in the lower energy and can be dominated by a retention mode at higher energies [3.1.30, 31].

In consistence with trajectory calculation an evidence of inversion process seems to be presented for T-for-F reaction in CH_3F by low energy tritium (2.7 eV) produced by photolytic decomposition of TBr [3.1.32].

Isotope Effects

In Table 3.1.8 [3.1.18] are listed the reactive isotope effects of hot tritium atoms with organic molecules. It is seen that for all the recoil tritium systems, the isotope effect is about 30% in favor of the protonated over the deuterated systems for both T-to-HT and T-for-H reactions. Of these molecules listed, only CHF_3/CDF_3 and $(CH_3)_3CH/(CH_3)_3CD$ pairs are suitable for the measurements of the pure primary isotope effects. The pure secondary isotope effect is also measured in T-for-F reaction in CH_3F/CD_3F and CHF_3/CDF_3 pairs as 1.40 ± 0.05 and 1.45 ± 0.05, respectively.

Isotope effects are also measured in reactions of hot hydrogen atoms produced from photolysis. In the reactions of 2.8 eV tritium with variously deuterated methanes (CH_4, CH_3D, CH_2D_2, and CD_4) [3.1.33] the HT/DT ratio per bond basis falls within the range of 1.4—2.0 for all molecules, inter- and intra-molecularly, indicating the importance of primary isotope effect.

As for the substitution reactions, the relative yields for CH_4, CH_3D, CH_2D_2 and CD_4 are 7.2 ± 0.2, 5.6 ± 0.3, 3.1 ± 0.3 and 1.0. The measurement of intramolecular isotope effect in the replacement of H vs. D, however, gives per bond substitution ratio with a value of approximately unity. These results are quite consistent with values of 1.6 ± 0.2 for both the primary and secondary isotope effects.

Martin et al. [3.1.34] have photolyzed the HBr—RD and DBr—RH systems with various wavelengths of light. By applying the steady state approximation to the reactions involved, the results are summarized to give the composite isotope effect $[D^* + RH]/[H^* + RD]$ for three initial hot atom energies. This composite isotope effect, favoring H-abstraction by D over D-abstraction by H (2.2 ± 0.2), is nearly independent of the structure and the initial energy over the range studied.

Table 3.1.9 shows the intramolecular isotope effects in the reactions of hot H and D with CH_3CD_3 [3.1.35]. In combining these results with those (1.6 ± 0.3) obtained for HT/DT ratio in the reaction at 2.8 eV with CH_4/CD_4 pair, the results are compatible with the trend that the isotope effect decreases with the increasing mass of hot atoms. These trends qualitatively agree with the results of the trajectory calculation [3.1.36].

Reaction Model

In the reactions of recoil tritium with alkanes, the HT/RT ratio increases from 0.79 for CH_4 to about 2.5 for compounds with higher complexity. Wolfgang

Table 3.1.8. Reactive isotope effects in reactions of recoil atoms with organic molecules (taken from Ref. 3.1.18)

Isotopic compounds	Bulk components in system	$\dfrac{k_{HT}}{k_{DT}}$	$\dfrac{k_{T\ominus\text{fore}H}}{k_{T\ominus\text{fore}D}}$
H_2/D_2	Mixture (scavenged)	2.7	2.7
	Mixture (unscavenged)	1.55 ± 0.06	1.55 ± 0.06
H_2/D_2	Mixture (2.8 eV T)	0.98 ± 0.03	0.98 ± 0.03
HD	HD (2.8 eV T)	0.71 ± 0.01	0.71 ± 0.01
CH_4/CD_4	Mixture	1.28 ± 0.02	—
CH_4/CD_4	n-C_4H_{10}	—	1.31 ± 0.05
CH_4/CD_4	Mixture (2.8 eV T)	—	7.2 ± 0.1
CH_2D_2	CH_2D_2	1.4	—
CH_2D_2	CH_2D_2	1.32 ± 0.01	—
CH_2D_2	CH_2D_2 (2.8 eV T)	—	1.1 ± 0.1
CHD_3	CHD_3 (2.8 eV T)	—	1.2 ± 0.3
CH_3F/CD_3F	c-C_4H_8	1.26 ± 0.05	1.33 ± 0.04
CH_3F/CD_3F	He	—	1.27 ± 0.04
CHF_3/CDF_3	CH_3Cl	—	$[1.35 \pm 0.05]$
CH_3CD_3	CH_3CD_3	1.34 ± 0.01	—
n-C_4H_{10}/n-C_4D_{10}	CH_4	—	1.21 ± 0.05
$(CH_3)_3CH/(CH_3)_3CD$	Individual	$[1.65 \pm 0.06]$	$[1.25 \pm 0.05]$
Isopropyl benzoate	Isopropyl benzoate		
(Methyl groups)		—	1.41
(Aromatic ring)		—	1.16

Table 3.1.9. Hot-atom abstraction isotope effects for H* and D* + CH_3CD_3[a] (taken from Ref. 3.1.35; Copyright 1976, American Chemical Society)

E_i, eV	H_2/HD	HD/D_2
1.1	2.4 ± 0.1	
1.6	2.4 ± 0.1	2.0 ± 0.1
2.0	2.6 ± 0.1	1.9 ± 0.1
2.9	2.8 ± 0.1	1.9 ± 0.1
2.9[b]	3.1 ± 0.1	2.8 ± 0.4

[a] Error ranges indicate the range of the experimental data.

[b] Yields in high-Br_2 systems.

et al. have proposed the "steric obstruction model" for such observations [3.1.37]. In this model, the large variation observed in the HT/RT ratios was mainly attributed to the change in RT but not in HT yields. Later experiments, however, clearly show the variation of HT yields for various alkanes, suggesting the failure of the steric model.

a) Hydrogen abstraction. A systematic comparison of the relative HT yields for various molecules has been carried out in the presence of a third molecule as the bulk component. Compounds chosen as the third molecule are C_2D_4 [3.1.38] or c-C_4F_8 [3.1.39], which serve as both energy moderator and radical scavenger. A good correlation exists between the strength of the C—H bond under attack and the HT yield per bond basis as shown in Fig. 3.1.6. However, fluoromethanes, CH_2F_2 and CHF_3, give higher yields than that expected from the C—H bond strength involved. The explanation for this is as follows: the abstraction reaction is completed on a time scale faster than the configurational rearrangement of the residual radicals. Thus, in the H-abstraction from CH_4, the residual CH_3 group cannot attian the equilibrium configuration (planar structure) before the completion of abstraction: the radical would be in somewhat excited state. On the other hand, the equilibrium configuration of both CHF_2 and CF_3 are not planar and the derived redicals are almost in the equilibrium, or probably unexcited state. Consequently the HT yield from hydrocarbons will become lower than what would be expected from the bond dissociation energy values.

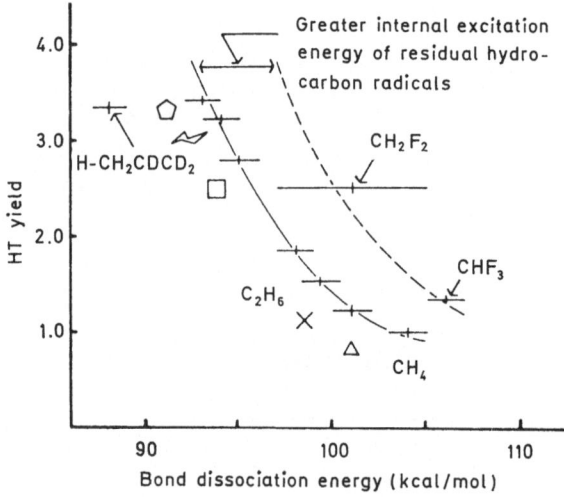

Fig. 3.1.6. Correlation with R-H bond dissociation energies of HT yield per bond from recoil tritium reactions with RH in perfluorocyclobutene moderator. (Reprinted with permission from Ref. 3.1.39. Copyright (1969) American Chemical Society)

Semi-quantitative estimation of the reaction time scale has been made based on the deviation of HT yield from $CH_3CD=CD_2$ molecule [3.1.40]. With a reasonable assumption a typical hydrogen abstraction reaction would take place in $2-5 \times 10^{-14}$ s.

Measurements [3.1.41] of per bond HT yields from various amines also indicate that the HT yield is inversely correlated with the bond strength of hydrogen atoms attacked.

Two models have been proposed to explain the mechanisms of hydrogen abstraction reaction: energy cut-off model [3.1.39] and stripping model [3.1.1]. In the former model, reactive collisions with C—H bonds can occur at lower energies for weaker C—H bonds, thereby increasing the total HT yield through a wider reaction range and a higher reaction probabilities over a substantial portion of the reaction energy range. Thus, this model is featured by lower-energy reaction of abstraction. The latter model is based on the completey different aspect and predicts the high-energy abstraction reaction which would be sensitive to the bond strength. When an energetic tritium passes by the C—H bond almost in perpendicular, a small fraction of excess energy is transferred to the C—H bond axis to induce abstraction of hydrogen. The results from hydrogen abstraction by photolytic tritium atoms indicate the preference of the lower-energy abstraction reactions. In the reaction with CH_4 of 2.8 eV tritium, the hot reaction yield is about 18%: the HT/CH_3T ratio is equal to 4. These results coupled with the fact that 50% of recoil tritium react through hot processes and HT/CH_3T is 0.79, reveal that low energy (2.8 eV) abstraction reactions may constitute an appreciable part of the observed nuclear recoil abstraction yields: 61% and 12% of the observed HT and CH_3T, respectively [3.1.42]. In another work with 2.8 eV tritium is also reported [3.1.43] the bond energy dependence of the abstraction reaction yield parallel to what has been observed with recoil tritium.

At present it is obvious that the energy cut-off model explains the observed bond strength effect for tritium atoms in several eV. However, there also exists an experimental evidence [3.1.44] for the stripping mechanisms at higher energies. Thus, it is likely that both reaction models are complementary in the recoil tritium reactions.

b) *Substitution reaction.* A good correlation is found between the T-for-H reaction yield and the NMR proton chemical shift for a variety of halocarbons and hydrocarbons [3.1.31]. Thus, it is obvious that the electronegativity instead of the steric factor is the parameter which governs the reaction predominantly: more electronegative substituents reduce the probability of T-for-H reactions to a larger extent.

The first mechanistic model (rotational inertia model) for the substitution reaction was proposed by Wolfgang et al. [3.1.1] to explain the decrease in the yield caused by introducing halogen atoms into CH_4. A critical test of the model has been attempted by measuring the yields of CH_2TX and CH_2TY from a molecule XCH_2CH_2Y, where X and Y are similar in mass but different in electronegativity. The results obtained with $CH_3CH_2CH_2F$ clearly limit the importance of the inertia effect as a controlling parameter [3.1.45].

It seems that physical and chemical factors may interact together in chemical reactions and it will be inadequate to discuss a single factor specifically without considering the other associated factors. Rowland et al. [3.1.33] have emphasized the importance of chemical interactions. Based on the results on the isotope effects on the reactions of 2.8 eV tritium with deuterated methanes, they have proposed the "pseudo complex" as the mechanistic model. Since it is likely that the replaced atom normally posses far less kinetic energy than the incoming tritium atom, the time scale for removing the replaced atom (H or D) to a distances of a least 1 A is probably several times longer than the time for tritium atom to approach. In the $T^* + CH_4$ system, five hydrogenic substituents are so close to the carbon atom that one should expect appreciable concentrations of electron density between each hydrogen atoms and the central carbon atom for a period of several C—H vibrations or more. Such fleeting coexistence of five carbon-hydrogen interactions is described as "pseudo complex". The lifetime of the complex is estimated to be $3-7 \times 10^{-14}$ s which is long enough for the secondary isotope effect to operate. This model differs from the previous "rotational inertia model" in that the reactions are concerted and the motion of the residual atoms are not uniformly coordinated.

3.1.4 Fluorine

Flourine is the most reactive among the halogen atoms. The chemistry of recoil flourine species is most conveniently studied with ^{18}F atoms generated by a variety of nuclear reactions. In general, the observed yield from substitution reactions is lower nearly by an order of magnitude with recoil ^{18}F than with recoil tritium. Such a difference has been attributed to the following postulated factors [3.1.46, 47]. In the first place, ^{18}F atoms, which are heavier than tritium atoms, have slower velocity for the same energy and the collision time would be longer for ^{18}F atoms. Accordingly, more translational energy can be transferred on collision, and molecules reacted with recoil ^{18}F are vibrationally excited to a much greater extent than those involved in the recoil T reactions. Thus the extentive decomposition of the primary products is commonly observed. In the second place, the lower reactivity of ^{18}F is due to the greater steric hindrance to ^{18}F approaching a bond as compared to the approach of T atoms.

Scavenger

Because of the high reactivity of thermal fluorine atoms, no trace ingredient can successfully compete with a major component for the reactions. Thus it is not clear yet whether a really effective "scavenger" exists for near-thermal ^{18}F atom systems.

In most experiments performed so far, unsaturated compounds have been used together with HI or H_2S as a hydrogen donor to the resulting radicals:

$$^{18}F + RC{=}CR \rightarrow RC^{18}F{-}CR \cdot \xrightarrow{HI} RC^{18}F{-}CRH. \qquad (3.1.42)$$

As for the relative efficiencies of various ethylene derivatives in trapping thermal ^{18}F, Rowland et al. have examined the reactions of near-thermal ^{18}F with fluorinated ethylenes [3.1.48]. In their experiments, ^{18}F atoms are moderated

Table 3.1.10. Intermolecular and intramolecular selectivity in ^{18}F atom addition to olefins in excess SF_6 (taken from Ref. 3.1.48; Copyright 1972, American Chemical Society)

Olefin	Normalized yield per olefinic carbon atom[a]		
	CH_2	CHF	CF_2
$CH_2{=}CH_2$	1.0		
$CHF{=}CH_2$	0.7 (0.8)	0.6	
$CF_2{=}CH_2$	0.8 (1.1)		0.2
trans-$CHF{=}CHF$		0.3	
$CHF{=}CF_2$		0.4	0.1
$CF_2{=}CF_2$			0.14 (0.2)

[a] Yield per carbon atom of the stabilized radical formed by addition of ^{18}F to the listed group, relative to CH_2 in ethylene as 1.0. Numbers in parentheses are the yields after approximate correction for decomposition of excited radicals.

in kinetic energy by collision with excess SF_6, with which ^{18}F is quite unreactive [3.1.47]. Most ^{18}F atoms have lost most or all of their kinetic energy, and react with minor components chiefly as thermal or near-thermal ^{18}F atoms. Thus, the observed reactions should be quite similar in relative yields to those in the studies of true thermal ^{18}F atoms. The results are summarized in Table 3.1.10. In intermolecular competition, the ^{18}F atoms react preferentially with the less fluorinated molecules. The ^{18}F addition is also preferred intramolecularly at the less fluorinated end of the molecule.

Abstraction

Root and his coworkers [3.1.49] have successfully measured the yields of the labile products $H^{18}F$ and $F^{18}F$. The Kimax glass of which the equipment was constructed served as a trap for $H^{18}F$, hydrogen fluoride. The $F^{18}F$ was collected on a potassium carbonate stripper. These experimental procedures enabled complete determination of all the recoil fluoride produced in the system. The results on the reactions with CH_3CF_3 are summarized in Table 3.1.11, and the most important reaction mode was abstraction of H by hot ^{18}F to form $H^{18}F$ (51%). Only about 17% of ^{18}F atoms reacted in thermal energies. In the absence of scavengers, these atoms also form $H^{18}F$.

The abstraction reaction by near-thermal ^{18}F from various protonated compounds such as H_2S, CH_4, D_2, C_2H_6 and CH_3CD_3 were also studied: SF_6 also constitutes a major component (90%) [3.1.50]. In the presence of acethylene as a minor component, the following reactions account for 95% of the ^{18}F atoms formed, i.e., H-abstraction (43), and addition (44) on acetylene:

$$^{18}F + C_2H_2 \rightarrow H^{18}F + C_2H, \tag{3.1.43}$$

$$^{18}F + C_2H_2 \rightarrow [CH^{18}F{=}CH]^* \xrightarrow{\text{HI or } H_2S} CH^{18}F{=}CH_2. \tag{3.1.44}$$

The inclusion of a fourth component, RH, permits additional reactions, especially, abstraction from this molecule. Although no direct analysis is usually performed for $H^{18}F$, the total yield of $H^{18}F$ was estimated indirectly from the diminution in the $CH_2{=}CH^{18}F$ yield, another major product from near-thermal ^{18}F atoms in these systems. The data, corrected for the reactions with acetylene and HI (or H_2S), show a good linear correlation with the RH/C_2H_2 ratio, from the slope of which the rate constant has been determined relative to that for the addition to acetylene.

Reactions with Perfluoriated Compounds

In the gas phase reactions of recoil ^{18}F atoms with CF_4, the $CF_3{}^{18}F$, products from ^{18}F-for-F reaction, will mainly decompose to $:CF^{18}F$ and partly to CF_3I (in the presence of I_2); the apparent yield of $CF_3{}^{18}F$ is about 4% of the recoil ^{18}F available for reactions. The total initial yield of this product is about 12—14% as compared with the yield of about 30% for CH_3T from the tritium reaction with CH_4 [3.1.51]. With $CF_2{=}CF_2$, recoil ^{18}F generates highly excited $CF_2{=}CF^{18}F$ by displacement, and $\cdot CF_2CF_2{}^{18}F$ by addition [3.1.52]. Both molecules are prone to decomposition and yields of the labeled parent molecules are quite small.

Table 3.1.11. Summary of primary yield and energetics data for CH_3CF_3 (Ref. 3.1.59)

Primary hot reaction	Primary hot yield (%)	Low-pressure yield fractions (%)		Excitation energy ranges (eV)	
		S/(S + D)	D/(S + D)	Thermochemical minimum	Kinetic average
F-to-HF abstraction	51 ± 3				
F-to-F$_2$ abstraction	5.4 ± 0.3				
F-for-F substitution	3.56 ± 0.07	0.15 ± 0.01	0.85 ± 0.02	3.0 ± 0.1	8.3 ± 1.5
F-for-F (D by β elimination)			0.85 ± 0.02	3.0 ± 0.1	8.2 ± 1.5
F-for-F (D by C-C scission)			0.00 ± 0.05	4.3 ± 0.1	
F-for-2F substitution	< 0.1				
F-for-H substitution	8.22 ± 0.09	0.24 ± 0.01	0.76 ± 0.02	3.0 ± 0.1	
F-for-H (D by β elimination)			0.043 ± 0.003	3.0 ± 0.1	
F-for-H (D by C-C scission)			0.23 ± 0.01	4.3 ≦ E ≦ 7.9	6.3 ± 1.0
F-for-H (D by C-C scission)			0.49 ± 0.02	7.9 ± 0.2	12 ± 5
F-for-2H substitution	< 0.1				
F-for-CH$_3$ replacement	5.79 ± 0.31	0.15 ± 0.01	0.85 ± 0.02	5.5 ± 0.1	
F-for-CH (D by C-F scission)			0.06 ± 0.01	5.5 ≦ E ≦ 9.3	
F-for-CH$_3$ (D by C-2F scission)			0.79 ± 0.02	9.3 ± 0.1	
F-for CF$_3$ replacement	8.5 ± 2.5	0.15 ± 0.02	0.85 ± 0.04	3.5 ± 0.2	
Total organic yield	26.1 ± 2.5	4.5 ± 0.1	21.6 ± 1.5		
Total hot yield	83 ± 3				

Other characteristic features are: (i) The recoil fluorine attack at C—C sites is about twice as efficient as at C—F site, in striking contrast to the preference for C—H position in recoil T reactions. (ii) When recoil ^{18}F reacts with cyclic perfluorocarbon molecules, the favored decomposition route of the excited primary species is via the fission of two C—C bonds to produce :CF$_2$ and C$_2$F$_4$ from c-C$_3$F$_6$ and :CF$_2$ and c-C$_3$F$_6$ from c-C$_4$F$_8$. The excited c-C$_4$F$_8$ also cleaves to produce 2C$_2$F$_4$ [3.1.53].

Reaction with Hydrocarbons

The reactivity of recoil ^{18}F toward CH$_4$ is roughly one-third of that toward CF$_4$ [3.1.54]. The decomposition of CH$_3{}^{18}$F to CH$_2$ + H^{18}F was much less marked than that of CF$_3{}^{18}$F.

In the reactions with cyclic hydrocarbons, about half of the c-C$_3$H$_5{}^{18}$F is converted to various isomeric fluoropropylenes at 100 torr and two thirds of c-C$_4$H$_7{}^{18}$F decomposes to vinyl fluoride. Vinyl fluoride is also a prominant product from both c-C$_3$H$_6$ and 1,3-dimethylcyclobutane. The decomposition of the F-for-CH$_3$ reaction product will be responsible for the formation of the latter.

When a recoil ^{18}F atom adds onto a double bond, the resulting excited radical can decompose not only by C—H rupture but by breaking C—C bond to give ·CH$_2{}^{18}$F and :CH$_2$. It is suggested that these decomposition reactions are characteristic of molecules in which the excess energy is to induce not only vibrational but also rotational excitation resulting from collisions with high impact parameters.

The isotope effect in the F-for-H or D reactions of hot ^{18}F with CH$_4$ and CD$_4$ is about 30% in favor with CH$_4$ over CD$_4$ [3.1.55].

Reactions with Fluorinated Compounds

In the reactions of recoil ^{18}F with fluoromethanes [3.1.51], both the ^{18}F-for-H and ^{18}F-for-F reactions in CH$_2$F$_2$, CHF$_3$ and CF$_4$ are accompanied by extensive decomposition of the primary products in the pressure range from 1 to 3 atm: the decomposition occurs in more than 80% of CH$_2$F^{18}F and CHF$_2{}^{18}$F, and in about 50% of CF$_3{}^{18}$F. The yields of carbenes, CH^{18}F and CF^{18}F (both in the singlet electronic states) are measured by trapping them with various olefines (ethylene, propylene, cis- or trans-2-butene, and isobutylene) to give the corresponding fluorocyclopropanes [3.1.51, 56], e.g.,

$$^{18}F + CH_2F_2 \rightarrow [CH_2F^{18}F]^* + F \rightarrow CH^{18}F + HF + F \qquad (3.1.45)$$

$$[CHF_2{}^{18}F]^* + H \rightarrow CF^{18}F + HF + H \qquad (3.1.46)$$

$$CH^{18}F \text{ or } CF^{18}F + RC{=}CR' \rightarrow RC{-\!-}CR' \text{ or } RC{-\!-}CR' \qquad (3.1.47)$$
$$\diagdown \! \diagup \qquad\quad \diagdown \! \diagup$$
$$CH^{18}F \qquad\qquad CF^{18}F$$

The addition reactions are not sufficiently exothermic to cause decomposition or isomerization of the fluorocyclopropanes formed. Ethylene was found to be unreactive toward CF^{18}F, while isobutylene scavenges the CF^{18}F quantitatively.

Root et al. have studied extensively the reactions of recoil ^{18}F with CH_3CF_3 over the fugacity range from 0.026 to 30 atm [3.1.49, 57—59]. The primary absolute yields obtained in these experiments are also shown in Table 3.1.11. The F-for-H reaction is favored over the F-for-F reaction. The primary alkyl replacement reactions in CH_3CF_3 were more probable than atom-for-atom substitution reactions by a per-bond factor of 7.3 ± 1.3. This is opposite to the systematics observed in recoil tritium reactions with alkanes, where the T-for-H substitution was roughly threefold more probable than the alkyl replacement [3.1.60].

Mechanisms and Energetics of Recoil ^{18}F

The ^{18}F-for-F reaction with both the dl- and meso-$(CHClF)_2$ in the gas phase also ends up with the retention of configuration [3.1.61].

As for the energetics of the reactions, the ^{18}F-for-H reaction product from CH_3CF_3 decomposes via both C—C bond fission and HF elimination [3.1.59]. However, no rupture of the C—C bond is observed in the ^{18}F-for-F reaction product. This may indicate that the energy is localized in various parts of the molecule, or that the excitation of rotational modes is particularly important [3.1.62].

The possible role of the bond strength in determining the yields of ^{18}F-labeled products has been investigated with a series of alkyl halides (RX). The primary yield of $R^{18}F$ increases as $D(R-X)$ decreases (X = F, Cl, Br, or I) by factor of 8 from CH_3F to CH_3I [3.1.63]. Other factors (both physical and chemical in nature) must be considered to generalize the observed trend. However, it appears that at least, substitution reactions initiated by ^{18}F atoms are influenced strongly by the chemical factors associated with the change in the C—X(C—H) bond strength.

An evidence for the bond energy dependence of the abstraction reactions has been reported [3.1.50]. The relative yields for abstraction from the saturated C—H bonds indicate a good correlation with the dissociation energy (in parenthesis, in kcal/mol) as is the case with the recoil tritium reactions: $C_2H_6(98) > CH_4(104) > CH_3CF_3(109)$. The formation of the 135 kcal/mol H—F bond makes all the abstraction highly exothermic; the activation energy barrier apparently increases slowly for the very strong C—H bonds. However the data are sparse for a detailed evaluation of the correlation between the abstraction yield and the C—H bond dissociation energy.

3.1.5 Chlorine

Chlorine atoms with much higher recoil energies are produced by nuclear reactions such as $^{35}Cl(\gamma, n)^{34}Cl^m$ and $^{40}Ar(\gamma, p)^{39}Cl$. The maximum recoil energy of Cl atom from the (n, γ) reactions is in several hundred eV and a small fraction of events may fail to lead to the rupture of original bonds. However, the experimental results on dichlorobutane indicate the upper limit to the failure of bond rupture as 0.02% for ^{38}Cl [3.1.64]. The subsequent hot reactions are essentially similar among those nuclides and can be assumed to involve neutral atomic chlorine with excess kinetic energy [3.1.68]. Thus, similar product distributions are obtained as long as an adequate scavenger for the thermalized Cl atoms is employed in an adequate quantity.

Various scavengers have been tested for their efficiencies. Bromine [3.1.65] and iodine [3.1.66] act efficiently as scavengers for thermal ^{38}Cl atoms. With Br_2 as the scavenger, the yields of hot reaction products in the gaseous $^{38}Cl-CH_4$ systems are: $CH_3^{38}Cl$, $3.4 \pm 0.25\%$; $CH_2Cl^{38}Cl$, $0.72 \pm 0.07\%$; $\cdot CH_2^{38}Cl$, $2.14 \pm 0.12\%$, and $CH^{38}Cl$, $< 0.3\%$ [3.1.65]. The $\cdot CH_2^{38}Cl$ radical is formed by the C—H or C—Cl bond cleavage following the displacement of both H and Cl atoms from CH_3Cl, and is detected as $CH_2^{38}ClBr$. The $CH^{38}Cl$ radical resulting from the HCl-elimination from $CH_2^{38}ClCl$, was stabilized as $CH^{38}ClBr_2$.

The extent of secondary unimolecular reactions subsequent to substitution reactions has been further examined for various reactant molecules [3.1.69]. In this systematic study, the complementary products of the secondary decomposition have been traced with systems in which the substitution products sometimes react via a pathway with survival of the C—Cl bond. Such products include dichloro compounds for which loss of the non-radioactive Cl atom can occur in 50% of such reactions, and chlorocyclanes which do not decompose through C—Cl bond cleavage.

The isotope effects in the recoil ^{38}Cl reactions with CH_4 and CD_4 have been measured by diluting the systems with an inert gas, Ar, to establish the steady state collision distribution [3.1.67]. The only primary products from hot reactions observed were $CH_3^{38}Cl$, and $\cdot CH_2^{38}Cl$ radical measured experimentally as $CH_2^{38}ClI$ after being scavenged by I_2. The ratios of products from CH_4 to those from CD_4 remain nearly constant (1.8 for $CH_3^{38}Cl/CD_3^{38}Cl$ and 1.6 for $CH_2^{38}Cl/CD_2^{38}Cl$) over the variation of the Ar mole fraction from 0.6 to 0.9.

In the reactions of ^{38}Cl with $H_2(D_2)$ in the presence of composite scavenger, ethylene and I_2 [3.1.70], the $H^{38}Cl(D^{38}Cl)$ results from the direct abstraction reaction of ^{38}Cl:

$$^{38}Cl + H_2(D_2) \rightarrow H(D)^{38}Cl + H(D), \qquad (3.1.48)$$

$$^{38}Cl + C_2H_4 \rightarrow H^{38}Cl + C_2H_3. \qquad (3.1.49)$$

The thermal chlorine atoms eventually result in the formation of $C_2H_4I^{38}Cl$:

$$^{38}Cl + C_2H_4 \rightarrow C_2H_4^{38}Cl \xrightarrow{I_2} CH_2ICH_2^{38}Cl. \qquad (3.1.50)$$

The $H^{38}Cl$ and $D^{38}Cl$ yields extrapolated to 1.0 mole fraction CF_2Cl_2 were ~ 10.6 and $\sim 8.3\%$, respectively. Subtraction of the $H^{38}Cl$ yield (6.5%) from the ethylene reaction results in the measured low energy isotope effect of $k_{48(H)}/k_{48(D)} = 2.3$. In thermally equilibrated systems, normal isotope effects have been determined experimentally as $k_{48(H)}/k_{48(D)} = 9.4$ at 300 K, for the $Cl + H_2(D_2)$ abstraction reactions, which decrease rapidly at higher temperatures to the limiting minimum value predicted as ~ 1.4. Thus, extrapolation of the temperature dependence of the thermal isotope effect gives an effective minimum reaction temperature of ~ 1200 K for the reactions at high moderation in the presence of 5% ethylene.

At low moderator concentrations (corresponding to high average energies of reaction), the $H^{38}Cl$ and $D^{38}Cl$ yields exhibit the inverse isotope effect, $Y_{DCl} > Y_{HCl}$. The net absolute yields from the chlorine atom reaction with H_2 and D_2 at 50% moderation were measured to be 16 and 26%, respectively, with the inverse isotope effect of 0.62.

The above $^{38}Cl + H_2(D_2)$ reactions were further examined by using non-Boltzmann kinetic theory in order to explore the high energy isotope effect. The rate equations with H_2 as reactant can be summarized as

$$[Y/(1 - Y)] = (k_{48(H)}/k_{50}) \left(\frac{[H_2]}{[C_2H_4]} \right) + (k_{49}/k_{50}) \tag{3.1.51}$$

when Y denotes the fractional yield of $H^{38}Cl$. The ratio of the reactant concentration plotted against the measured yield term, $[Y/(1 - Y)]$, can be approximated by the linear correlation over the range of reactant concentration from zero to 66%. At lower moderations (corresponding to higher average energies of reaction) the correlation is significantly deviated from linearity owing to the breakdown of the time-independent steady state assumption in the highly reactive hydrogen system, and to the difference in the energy dependence of the individual reaction monitored. Extrapolation of the linear region of the curve to infinite moderation gives an intercept of $k_{49}/k_{50} = 0.08$. Thus $k_{50} = 12k_{49}$ and Eq. (3.1.51) can be further simplified as

$$\frac{k_{49}}{k_{48(H)}} = \frac{(1 - Y) [H_2]}{(13Y - 1) [C_2H_4]}. \tag{3.1.52}$$

In Fig. 3.1.7, the ratios of $k_{49}/k_{48(H)}$ and $k_{49}/k_{48(D)}$ calculated from Eq. (3.1.52) for various sample compositions, were plotted as a function of the moderator mole fraction. Considering the factors which may account for the observed results it follows that the crossover in the isotope effect reflected in the rate constant ratio k_{48}/k_{49} is kinetic in nature.

The substitution at asymmetric carbon by recoil chlorine (^{38}Cl from (n, γ) and ^{39}Cl from (γ, p)) also proceeds with almost complete retention of the optical configuration in the gaseous dl- or meso-2,3-dichlorobutane [3.1.64]. An evidence

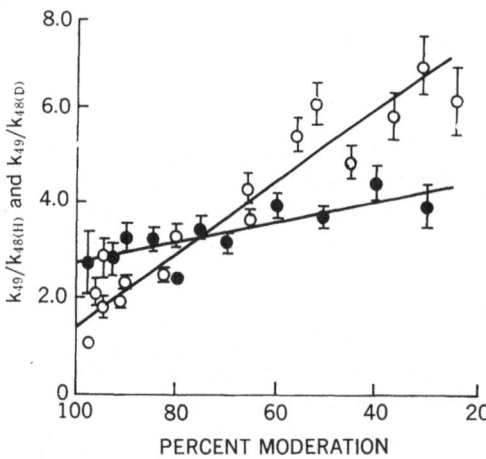

Fig. 3.1.7. Calculated non-Boltzmann rate constants $k_{49}/k_{48(H)}$ and $k_{49}/k_{48(D)}$ as a function of moderation with CF_2Cl_2. $\circ$, $^{38}Cl + H_2$ reaction; $\bullet$, $^{38}Cl + D_2$ reaction. Error bars represent the standard deviation. (Reprinted with permission from Ref. 3.1.70. Copyright (1978) American Chemical Society)

for the ^{38,34m}Cl-for-Cl substitution reaction with inversion has been reported for 2(S)-chloropropionyl chloride and 2(R)-chloropropionyl chloride in gaseous and condensed phase [3.1.71]. The retention-to-inversion ratio of optical configuration at the reaction was $0.21 \sim 0.6$ in the gas phase and 1.13 ± 0.02 in the liquid phase. There exist three rotational conformations of 2-chloropropionyl chloride: trans, gauche and gauche prime forms. In the gaseous state, the gauche prime conformation becomes significantly more abundant than in the liquid phase, while only the trans form persists in the solid state. The gauche prime conformation offers a relatively unhindered approach to attack of the asymmetric carbon from the backside (with respect to the 2-chlorine atom). Thus if the gauche prime conformation is the open conformation which determines gas phase Cl-for-Cl substitution with such a high degree of inversion, blocking the backside with large substituent groups should hinder the inversion mode. The assumption that the degree of retention (or inversion) should be strongly affected by steric hindrance is consistent with the results from 2(S)-chloro-4-methylvaleryl chloride in the gaseous phase, in which the backside attack is sterically hindered and the percentage retention of ^{34m}Cl-for-Cl substitution at the asymmetric carbon is $59.3 \pm 0.8\%$ (48.6 ± 1.3 in the liquid phase). Thus, it appears that not all hot substitution reactions are fast and direct processes involving front-side approach alone, but can also occur from the backside of the molecule, resulting in Walden inversion.

3.1.6 Bromine

The two naturally occurring isotopes of bromine, ^{79}Br and ^{81}Br, provide radioactive nuclides through (n, γ) process. In both cases, metastable nuclei are also produced which decay by isomeric transition (I.T.). Other nuclear reactions which yield radiobromine are ^{79}Br(n, 2n)^{78}Br and ^{81}Br(n, 2n)^{80m}Br, both induced by 14 MeV neutrons. Recoil bromine atoms (^{84}Br, ^{86}Br and ^{87}Br) are also produced by uranium fission, and have been found to react with methane to produce methyl bromide.

The (n, γ) process results in the energy distribution ranging from a very low (near zero) value to $\sim 10^2$ eV at maximum. The reactions of various Br atoms from (n, γ) processes with CH_3F and CD_3F have been examined at varying Br_2 concentrations. The limiting organic yield at zero mole fraction (m.f.) of Br_2 decreases in the order: ^{80m}Br $> ^{80}$Br $> ^{82+82m}$Br [3.1.72]. By adding He or Ar as moderator, the organic yield decreases and can be extrapolated to zero at 100% moderation: this indicates that the reaction occur entirely as a result of the recoil kinetic energy acquired by the bromine in their activation processes. In the reactions of (n, γ)-generated ^{80}Br with CH_4 and CD_4 in the presence of small amounts of Br_2, kinetic energy independent (thermal ionic) processes also contribute to the observed yields (CH_3^{80}Br, 12.4%; CH_2^{80}BrBr, 1.5%; CD_3^{80}Br, 4.7%; CD_2^{80}BrBr, 1.8%) [3.1.73]. The complex phenomenon is primarily due to the charges imparted to ^{80}Br at the time of nuclear reaction. Because of the relatively low ionization potential (I.P.) of bromine, certain fractions of their ions can be thermalized without being neutralized, and react through ion-molecule reactions which can be blocked neither by radical scavengers, nor by inert gases as moderators.

The isomeric transitions of both ^{80m}Br and ^{82m}Br are highly internally converted, followed by the Auger processes, and the molecules will explode. The kinetic energies and charges imparted to the atoms in such events are greatly dependent on the remaining atoms in the parent molecules. These energetic ions will be neutralized easily to $+1$ state at an early stage. However, further neutralization is usually endothermic and less likely to occur.

In the reactions of ^{80}Br with C_2H_6 using $^{80m}BrBr$ as the source material, the products were $C_2H_5{}^{80}Br$ (1.8%), $CH_2{}^{80}BrBr$ (0.8%) and $CH_3{}^{80}Br$ (2%) [3.1.74]. The reactive species leading to the formation of the first two products is in fact the hot atom in its electronic ground state. The yields from C_2D_6 were in all cases lower by about 30% than from C_2H_6. Methyl bromide can be formed via both hot and ionic processes, and that the latter becomes dominant as the amount of moderator is increased. In the reactions of ^{80}Br from $^{80m}BrBr$ with C_2H_6, C_3H_8, and CH_3Br (I.P.s are 11.65, 11.21, and 10.6 eV, respectively), the products are essentially formed via energetic processes [3.1.75, 76].

Table 3.1.12. Percentage yields of organic products due to kinetic and thermal-ionic processes in the I.T. and (n, γ)-experiments (HBr (or Br_2)/$CH_4 = 0.1 \pm 0.01$) (from Ref. 3.1.77)

System	$H^{82m}Br—CH_4$	$H^{80m}Br—CH_4$	$Br^{82m}Br—CH_4$	$Br^{80m}Br—CH_4$	(n, γ)-activated $^{82m}Br_2$ $—CH_4$
E_{max} (eV)	1.3	1.3	158	158	417
Org. yield (%)	4.5	4.25	6.1	4.7	13.8
Kinetic process:					
$\quad CH_3Br$ (%)	0	0	4.5	3.0	11.4
$\quad CH_2Br_2$ (%)	0	0	0	0	1.0
Thermal-ionic:					
$\quad CH_3Br$ (%)	0.8	1.60	0.5	0.5	0.9
$\quad CH_2Br_2$ (%)	3.7	2.65	1.1	1.1	0.5

When $H^{80m}Br$ or $H^{82m}Br$ is used as source molecules for recoil ^{80}Br or ^{82}Br, thermal ionic reactions becomes much more important [3.1.77]. In the reactions with CH_4, essentially all the products (mainly CH_3Br and CH_2BrBr) are formed via kinetic-energy independent processes, and their yields are not influenced by the addition of additives except at higher moderation [3.1.77]. In Table 3.1.12, various recoil Br reactions with CH_4 are summarized. Distinct isotope effects have also been found between I.T.-generated ^{80}Br and ^{82}Br. This isotope effect will be ascribed to the difference in the decay scheme between ^{80m}Br and ^{82m}Br. Although the M3 transitions in ^{80m}Br and ^{82m}Br are essentially similar, the excited ^{80}Br has a subsequent transition to the ground state. Sixty-one percent of the excited ^{80}Br is deexcited via the second internal conversion process, while the remaining 39% falls to the ground state by emission of 37 keV γ-rays. The chemical consequences resulting from this γ-emission are considered as negligible since the kinetic energy acquired by the ^{80}Br is in the order of magnitude of 10^{-2} eV, which is not large enough to break any bond in the molecule. Accordingly, the reaction processes of I.T.-generated ^{80}Br can be classified into the following two types:

Process A (39%): internal conversion, and 37 keV γ-ray emission (no chemical effects).

Process B (61%): internal conversion, and internal conversion.

The analysis of the experimental results shows that there exists a striking difference between the yield distributions for individual products from the Processes A and B; it is evident that Process B is responsible for such isotope effect.

As for unimolecular decomposition of primary excited products, only one type of compound, [c-$C_3H_5{}^{80}$Br], has been investigated [3.1.78] with (n, γ)-activated ^{80}Br. The decomposition pathways of the primary product formed either by ^{80}Br-for-H in c-C_3H_6 or by ^{80}Br-for-Br in c-C_3H_5Br are

$$[\text{c-}C_3H_5{}^{80}Br]^* \rightarrow (CH_2CHCH_2{}^{80}Br)^*, \qquad (3.1.53)$$

$$(CH_2CHCH_2{}^{80}Br)^* \rightarrow CH_2CH + CH_2{}^{80}Br, \qquad (3.1.54)$$

$$\rightarrow CH_2CHCH_2 + {}^{80}Br. \qquad (3.1.55)$$

Figure 3.1.8 shows the potential energy profiles for these three reactions, together with the activation energies. In the c-C_3H_6 system, the majority of the c-$C_3H_5{}^{80}$Br (~80%) resulting from the ^{80}Br-for-H reaction is internally excited by as much as 90 kcal/mol, decomposing to $CH_2{}^{80}$Br. On the other hand, in the c-C_3H_5Br system, [c-$C_3H_5{}^{80}$Br]* will mainly end up as $CH_2CHCH_2{}^{80}$Br, or as allyl radical and ^{80}Br atom; the yield of $CH_2CHCH_2{}^{80}$Br is more than 80% of the total organic yield at 100 Torr. The results imply that the excitation energy of the primary product from the ^{80}Br-for-Br reaction is between 47.3 and 90 kcal/mol (2—4 eV).

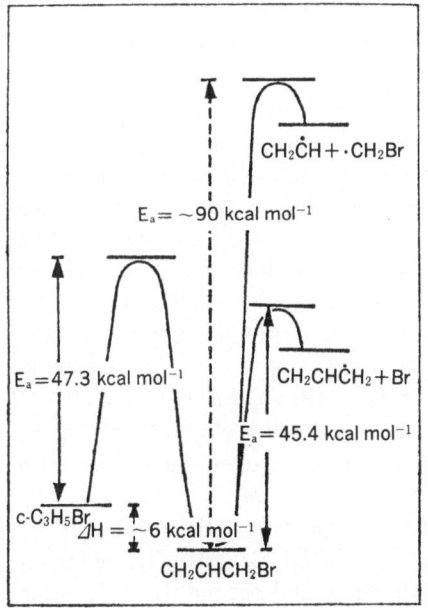

Fig. 3.1.8. Potential energy diagram of reactions (3.1.53), (3.1.54) and (3.1.55) (taken from Ref. 3.1.78)

3.1.7 Iodine

A variety of reactions can be utilized to produce recoil iodine: they include
$^{127}I(n, \gamma)^{128}I$, $^{127}I(n, 2n)^{126}I$, $^{129}I(n, \gamma)^{130}I$ and $U(n, f)^{123 \cdot 125}I$. Decays of the radioactive precursors such as Te and Xe, also give radioiodine with a suitable half-life.
However, the chemistry of recoil iodine is complex. The (n, γ)-activated iodine
can react as ions or atoms, hot or thermal, electronically excited, or in combination
of these qualities depending upon the reaction system. Furthermore, it sometimes
appears, that radiation-induced reactions should also be invoked so as to account
for the experimental results. In general, low yields for hot reactions are attributed
to the large mass of iodine atom; it moves to collide with a molecule so slowly,
even when hot, that the molecule has a chance to relax vibrationally in the
meantime.

Rack and Gordus [3.1.79] measured the organic yields from the (n, γ)-activated
^{128}I reaction with CH_4 as a function of the mole fraction of inert gases in the
presence of 0.5 mmHg CH_3I and 0.1 mmHg I_2. The data extrapolated to zero
m.f. methane indicate that Xe is capable of reducing the organic ^{128}I yield to
11%, whereas Ne, Ar, and Kr each reduces it to about 36%. It may be concluded
that out of the 54.4% organic ^{128}I, about 18.4% is formed through hot ^{128}I
reaction, 11% via reactions of excited iodine atoms or I^+ ($^3P_2$, $^3P_1$, and $^3P_0$)
ions, and 25% via reactions of $I^+(^1D_2)$ ions. Thermochemical consideration
on the formation of CH_3I from the iodine in various excited state and CH_4,
however, predicts that the reaction yielding CH_3I^+ or CH_3I is exothermic only
with $I^+(^1D_2)$. Hence, Loberg and Welch [3.1.80] have examined the ionic reactions
of ^{123}I from ^{123}Xe, which decays in two modes, i.e., electron capture and β^+
decay: in the latter decay mode 14% of the ^{123}I recoils are formed initially in
the I^- state. The proposed reaction model involves molecular ion complexes:
in the presence of additives, it is postulated that the reactive species is a molecular
ion, AI^+, where A is CH_4, Xe, Ne, Ar, Kr, or N_2. This species will undergo the
ion-molecular reaction to form organic products.

The ^{125}I obtained from $^{125}Xe \xrightarrow{EC} {}^{125}I$ process is preferable for the specific investigation of iodine ions, since the nuclide, ^{125}Xe, decays only via EC-process
and the resulting ^{125}I has a maximum kinetic energy of 15.6 eV. In the reactions
with gaseous CH_4, $CH_3{}^{125}I$ is the only organic product observed, whose yield
is very sensitive to the presence of electron scavengers such as I_2, SF_6, and O_2.
In Table 3.1.13 [3.1.81] are summarized the results obtained with various recoil
iodines. A general trend exists regarding the reactivities of excited iodine: hot
yields indicate remarkable dependence on the initial kinetic energy acquired
by the recoil iodine, and decrease in the order $(n, \gamma) >$ IT $>$ EC. The maximum
kinetic energies imparted to the recoil atoms are 194 eV for ^{128}I, and 177 eV
for ^{130}I, in (n, γ) process. In contrast, the thermal ionic yield was the highest
for ^{125}I obtained from the EC process, and the population of the iodine ion in
each excited state depends on the decay mode of precursors.

At the total pressure of 1 atm, the organic ^{128}I yield in C_2H_6 is $1.1 \pm 0.3\%$,
0.06% for CH_3I, 0.57% for C_2H_5I, and 0.46% for C_2H_3I [3.1.86]. The ionization
potential for C_2H_6 (11.65 eV) lies between the first and second I.P. of iodine,

Table 3.1.13. Comparison of the results obtained with various recoil iodines (from Ref. 3.1.81)

Reaction system	Iodine carrier	Yield of CH_3I (%)			
		Total	Hot	$^1D_2$	^3P
$^{127}I(n, \gamma)^{128}I + CH_4$	yes	54.4 ± 2.0	18.4 ± 2.0	25.0 ± 2.0	11.0 ± 2.0
$^{129}I(n, \gamma)^{130,130m}I + CH_4$	yes	42.5 ± 2.0	16.5 ± 2.0	9.5 ± 2.0	16.3 ± 2.0
$^{130m}I(IT)^{130}I + CH_4$	yes	25.6 ± 2.0	9.7 ± 2.0	5.6 ± 2.0	10.3 ± 2.0
$^{129}I(n, \gamma)^{130,130m}I + CD_4$	yes	41.3 ± 2.0	15.3 ± 2.0	9.5 ± 2.0	16.5 ± 2.0
$^{130m}I(IT)^{130}I + CD_4$	yes	26.4 ± 2.0	10.5 ± 2.0	5.6 ± 2.0	10.3 ± 2.0
$^{123}Xe(EC, \beta^+)^{123}I + CH_4$	no	51.8 ± 2.9	0	21.2	30.6
$^{123}Xe(EC, \beta^+)^{123}I + CH_4$	yes	53.8	0	27.0	25.4
$^{125}Xe(EC)^{125}I + CH_4$	—	58 ± 6	0	40 ± 7	18 ± 2^a
$^{125}Xe(EC)^{125}I + CH_4$	yes	75	0	55	20
$^{125}Xe(EC)^{125}I + CH_4$	yes	76.0 ± 6.0	8.7 ± 6.7	35.8 ± 8.0	31.5 ± 2.0

[a] Assigned as the yield resulting from the reaction of I^+ in the $^1S_0$ state.

and the charges distributed are rapidly reduced until the $+1$ species predominates. For a further neutralization of I^+,

$$I^+ + C_2H_6 \rightarrow C_2H_6^+ + I \qquad (3.1.56)$$

the heat of reaction is exoergic by 0.51 eV for $I^+(^1D_2)$ but over 1 eV endoergic for ground state I^+. This reaction serves as a quencher for $I^+(^1D_2)$.

Direct formation of an inorganic product (HI)

$$I^+ + C_2H_6 \rightarrow C_2H_5^+ + HI \qquad (3.1.57)$$

is exoergic by 0.93 eV for ground state I^+ and greater in excited states.

Ion-molecule reactions of the type

$$I^+ + C_2H_6 \rightarrow C_2H_5I^+ + H \qquad (3.1.58)$$

are mildly endoergic (0.8 eV) for $I^+(^3P_2)$ and exoergic for I^+ in excited states. However, the subsequent neutralization would impart internal energies (8.7 eV) leading to the decomposition of C_2H_5I. The formation of an excited complex of the type

$$I^+ + C_2H_6 \rightarrow [C_2H_6I^+]_{ex} \qquad (3.1.59)$$

has been postulated [3.1.86]. The neutralization will proceed by ejecting H^+ with a large portion of the energy. The reactions (3.1.56), (3.1.57) and possibly (3.1.59) may be generalized for all alkanes, and are exothermic except for CH_4 systems.

In the reactions of ^{128}I with n-butane [3.1.83] the hot reaction yield is only 1%, which consists of CH_2I_2 (31%), CH_3I (22%), C_2H_5I (21%), C_2H_3I (13%), n-C_3H_7I (9%) and sec-$C_5H_{11}I$ (4%). The absence of butyl iodide and the large yields of lower iodides reveal that the primary reaction products are produced with a great deal of vibrational excitation energies. In the liquid phase, however, n-butyl iodide is found with 26% yield. This is ascribed to rapid de-excitation of [n-$C_4H_9{}^{128}I$]* and also to caged recombination reactions involving thermalized ^{128}I and butyl radicals.

With ethylene and propylene as reactant gases [3.1.84], total organic yields were found to be 18 ± 1 and $24 \pm 2\%$, respectively. In both systems, $CH_3{}^{128}I$ was the major product. The important reaction involved is the ion-molecule reaction,

$$^{128}I^+({}^3P_0 \text{ or } {}^3P_1) + C_2H_4 \rightarrow [C_2H_4{}^{128}I]^+, \tag{3.1.60}$$

$$[C_2H_4{}^{128}I]^+ + C_2H_4 \rightarrow CH_3{}^{128}I + C_3H_5{}^+. \tag{3.1.61}$$

Unlike the reactions between the (n, γ)-activated ^{128}I and hydrocarbons, recoil reactions with halomethanes essentially consist in energetic hydrogen and halogen substitutions [3.1.85]. This is because of the lower ionization potentials of the reactant molecules, and the neutralization of $^{128}I^+$ ions is always exothermic. The limiting ^{128}I organic yield in gaseous CH_3F, CH_3Cl, CH_3Br and CH_3I progressively decreases in this order, as seen in Table 3.1.14. The only systematic trend observed was between the hot ^{128}I organic yield and the energy degradation factor in the halomethane system. The energy degradation factor depends on the masses of the hot atom and the colliding partner, and will not be important for atoms born with large kinetic energies such as (n, p)-activated tritium, but will be important for atoms or ions with energies in or near the reaction zone.

Table 3.1.14. Observed hot organic yields of (n, γ)-activated ^{128}I with halomethanes and the energy degradation factors calculated for these halomethane systems[a] (from Ref. 3.1.85; Copyright 1972, American Chemical Society)

Target molecule	CH_4	CH_3F	CH_3Cl	CH_3Br	CH_3I
% obsd. hot organic yields (reactions with (n, γ)-activated ^{128}I)	19.00 ± 2.0	11.20 ± 1.0	4.12 ± 0.5	0.67 ± 0.1	0.20 ± 0.1
Energy degradation factor	0.40	0.66	0.81	0.98	0.99
C—X bond energy, kcal/mol	104	108	884	70	56
% calcd. hot yields	19.00	13.43	13.43	13.43	13.43
IP, eV	12.99	12.8	11.22	10.54	9.54

[a] Bond energy, calculated hot yields, and IP have been added for the purpose of comparison.

Since no evidence for the reactant-isotope effect between CH_4 and CD_4 has been obtained with either I.T.-activated ^{130}I [3.1.82] or EC-generated ^{125}I [3.1.81], the isotope effect, if it exists at all, should be very small.

3.1.8 Carbon

Carbon-14 is produced by $^{14}N(n, p)^{14}C$ reaction, but long irradiation periods required inevitably produce appreciable radiation damages on both reactants and products. Therefore ^{11}C has been favored in the works of recoil carbon chemistry. Carbon atom is polyvalent and the reactions are characteristic and different from those of monovalent atoms. Carbon-11 is quite electron deficient, at its birth, and will rapidly tend to form chemical combination in which it shares more than four electrons. In the following are shown the most important primary reactions of ^{11}C:

(i) Insertion into the C—H bond

$$^{11}C + CH_3-CH_2-H \rightarrow CH_3-CH_2-^{11}C-H. \qquad (3.1.62)$$

(ii) Attack at the π bond in unsaturated compounds

$$^{11}C + CH_2=CH_2 \rightarrow H_2C \underset{^{11}\dot{C}}{\overset{\diagdown \diagup}{\quad}} CH_2. \qquad (3.1.63)$$

(iii) Reactions whose net reactions consist in abstraction of H atoms

$$^{11}C + CH_2-CH_3 \rightarrow -CH_2-CH_2 + ^{11}CH. \qquad (3.1.64)$$

Reaction with Alkanes

In the reactions with hydrocarbon molecules (RH), carbon atoms always give large yields of acetylene; ethylene and RCH_3 are also formed. The $[CH_3-CH_2{}^{11}CH]^*$ in Eq. (3.1.62) would be excited and its spin state would depend on whether the initial carbon has been ^1S, ^1D or ^3P. In the last case, the spins of the two electrons of the diradical would be parallel and the fragmentation might take place as follows:

$$[CH_3-CH_2-^{11}\dot{C}-H] \rightarrow [\dot{C}H_2=^{11}CH] + \cdot CH_3 \rightarrow$$
$$\rightarrow HC\equiv^{11}CH + H + \cdot CH_3 \qquad (3.1.65)$$

in which the weaker C—C bond has been supposed to break first. When ^{11}C is allowed to react with specifically deuterated and perdeuterated hydrocarbons and alkylfluorides [3.1.87], the major portion of acetylene produced is either $H^{11}C\equiv CH$ or $D^{11}C\equiv CD$. This is consistent with that carbon atom attack leading to the formation of acetylene is mainly an intramolecular process as Eq. (3.1.65). Further studies [3.1.88] on equimolar mixtures of hydrocarbons with their perdeuterated counterparts, however, indicate the presence of a minor intermolecular pathway also leading to acetylene.

Detailed studies on the yield variation for the reactions with hydrocarbons show that ^{11}CO decreased with increasing pressure and yields of $H^{11}C\equiv CH$ and $H_2{}^{11}C=CH_2$ increased with increasing pressure [3.1.89]. The "insertion and decomposition" model [Eqs. (3.1.62) and (3.1.65)] has been modified to accomodate the observed results. The idea is that the pressure trends observed can be accomodated by a model involving a collision complex as the first product for the interaction between the energetic carbon atom and the substrate.

Methyne has been proposed as the species most likely to give rise to ethylene rom a ny hydrocarbon (methyne insertion-decomposition mechanism):

$$R-H + {}^{11}C \rightarrow {}^{11}CH + R, \qquad (3.1.66)$$

$$R'-CH_3 + {}^{11}CH \rightarrow [R'-CH_2-{}^{11}CH_2], \qquad (3.1.67)$$

$$[R'-CH_2-{}^{11}CH_2] \rightarrow R' + CH_2={}^{11}CH_2 \qquad (3.1.68)$$

$$\searrow$$

"other products".

An alternative mechanism (vinyl radical mechanism), however, has also been proposed on the basis of the experimental results that ethylene-^{11}C formation is directly related with the number of methyl group in the substrate molecule [3.1.90].

Table 3.1.15. Relative and total ethylene-^{11}C yields from 1:1 mixture of protonated and perdeuterated hydrocarbons, and specially deuterated hydrocarbons[a,b] (Data corrected for isotropic composition of substrate) (taken from Ref. 3.1.91; Copyright 1975, American Chemical Society)

Substrate	% of total ethylene-^{11}C					Absolute yields[c] of ethylene-^{11}C
	${}^{11}C_2H_4$	${}^{11}C_2H_3D$	${}^{11}C_2H_2D_2$	${}^{11}C_2HD_3$	${}^{11}C_2D_4$	
$C_2H_6 + C_2D_6$	24.2 ± 0.8	33.9 ± 0.6	0.9 ± 0.7	19.4 ± 0.9	21.3 ± 0.5	16.6 ± 2.3
CH_3CD_3	23.4 ± 1.8	32.9 ± 1.3	1.9 ± 0.7	19.3 ± 0.9	22.5 ± 1.7	17.1 ± 1.0
$C_3H_8 + C_3D_8$	25.3 ± 1.7	29.2 ± 0.8	2.8 ± 1.2	19.8 ± 1.6	23.0 ± 1.6	14.1 ± 1.0
$CH_3CD_2CH_3$	68.2 ± 0.5	32.0 ± 1.1				13.7 ± 2.0
$CD_3CH_2CD_3$		0.7 ± 0.6	0.4 ± 0.6	26.8 ± 1.3	72.1 ± 1.0	13.2 ± 1.2
$CH_3CH_2CD_3$	34.0 ± 1.8	23.8 ± 0.5	1.3 ± 1.0	26.1 ± 0.9	14.8 ± 0.5	13.8 ± 0.9

[a] Data represent the arithmetic mean of results from three to eight determinations.
[b] Errors listed represent one standard deviation.
[c] As percent of total gaseous activity.

In order to test these reaction mechanisms, ^{11}C atoms were allowed to react with specifically deuterated hydrocarbons containing completely deuterated methyl or methylene groups and with equimolecular mixtures of protonated and perdeuterated hydrocarbons [3.1.91, 92]. The results (Table 3.1.15) suggest that the vinyl radical mechanism is not operative, or is only a minor pathway to ethylene formation. The results in the table further reveal that the ethylene formation exhibits two isotope effects, one in the formation of methyne (^{11}CH or ^{11}CD), another in the insertion step (^{11}CH or ^{11}CD insertion into a C—H or C—D bond), i.e., 1.2 in favor of ^{11}CD formation, and 1.4 in favor of methyne insertion into a CH_3 group, respectively.

Reactions with Alkenes

In the reactions of recoil ^{11}C atoms with alkenes, products from insertion and addition to the double bond are observed. With ethylene as reactant, sixteen products have been identified, although most of them are minor [3.1.93, 94]. The yields of major products at 1 atm ethylene are: 38.5% acetylene-^{11}C, 16.5% allene-^{11}C, 4.5% propyne-^{11}C, and 6.6% 1-pentyne-^{11}C [3.1.93]. The products fall into two groups of different origin: i) C_3 compounds, allene and 1-propyne and ii) acetylene and various unsaturated C_5 compounds (and small yields of C_4 compounds). Most allene is center-labeled ($CH_2=^{11}C=CH_2$) under any condition. End-labeling is more important in 1-propyne than in allene. They are formed by carbon insertion into ethylene π-bond.

The proposed reaction model for the production of acetylene-^{11}C involves the insertion of triplet carbon, $^{11}C(^3P)$, into the $C-H$ or $C=C$ bond of ethylene.

C—H insertion
$$^{11}:\dot{C} + CH_2=CH_2 \rightarrow CH_2=CH-^{11}\dot{C}H \leftrightarrow \dot{C}H_2-CH=^{11}\dot{C}H \tag{3.1.69}$$

$$\dot{C}H_2-CH=^{11}\dot{C}H \rightarrow \dot{C}H_2 + HC\equiv^{11}CH \tag{3.1.70}$$

C=C insertion
$$^{11}:\dot{C} + CH_2=CH_2 \rightarrow H_2C\text{---}CH_2 \rightarrow H_2C=^{11}C\dot{C}H_2 \tag{3.1.71}$$
$$\underset{^{11}\dot{C}}{\diagdown\diagup}$$

$$CH_2=^{11}C-\dot{C}H_2 \rightarrow \dot{C}H_2 + {}^{11}\ddot{C}=CH_2 \tag{3.1.72}$$

$$^{11}\ddot{C}-CH_2 \rightarrow H^{11}C\equiv CH \tag{3.1.73}$$

If decomposition to a stable molecule does not occur prior to collisional energy deactivation, the residuals will add ethylene to yield eventually certain specific C_5 products. The possible initial adduct of ^{11}C with ethylene can be summarized as shown in Fig. 3.1.9 [3.1.94].

When ethylene is moderated with heavy inert gases, Kr or Xe, the acetylene-$^{11}C/$allene-^{11}C ratio becomes higher as the molecular weight of the inert gas increases. This is because Xe, and to a lesser extent Kr, facilitates spin-orbital relaxation and so increases the population of ground-state $^{11}C(^3P)$ atoms. This would increase the acetylene-^{11}C yield at the expense of allene-^{11}C.

Reactions with Inorganic Molecules

When the π-bonded inorganic molecules [3.1.95], such as O_2, CO, CO_2 and SO_2, are reactants, ^{11}CO is the only significant product (yields $\sim 97\%$). Two models for attack of a ^{11}C atom on these inorganic molecules seem plausible. The first assumes the attack at the π-bond to form a cyclic intermediate, which can either be rearranged to $^{11}CO_2$, or decompose directly to ^{11}CO and O. The second possible mode of attack involves the "end-on" approach by the carbon atom to form a linear structural intermediate (e.g., $^{11}C-O-O$):

$$^{11}C + O=O \rightarrow [^{11}C-O-O]^* \rightarrow {}^{11}CO + O,$$

$$\Delta H = -138 \text{ kcal/mol}. \tag{3.1.74}$$

Such an intermediate can decompose to ^{11}CO, but formation of $^{11}CO_2$ is unlikely even in the condensed phase. The failure to observe $^{11}CO_2$ formation in the reaction of ^{11}C atom with liquid oxygen indicates that the "end-on" attack mechanism is dominant. An examination of the molecular orbitals involved indicates that this should in general be the preferred mechanism for ^{11}C atom attack on π-bonded inorganic molecules. With N_2, N_2O, and NO, ^{11}C atom can bond form a strong at either end of the molecule. In accordance with this, both ^{11}CO and ^{11}CN are observed as major products. With NO_2, ^{11}CO is the major product, again as predicted by the "end-on" attack mechanism. Both hot and thermal ^{11}C can undergo the exoergic reactions with O_2 to form ^{11}CO, and with N_2O to give ^{11}CO and ^{11}CN. By contrast, the endoergic process in which ^{11}C combined with N_2 to form ^{11}CN occurs only with hot atoms.

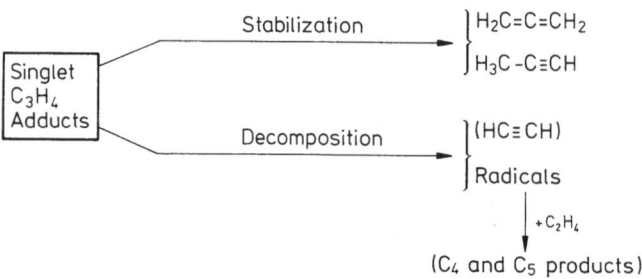

Addition of C (^{1}D) to ethylene. (Minor pathways are indicated by parenthesizing associated products.)

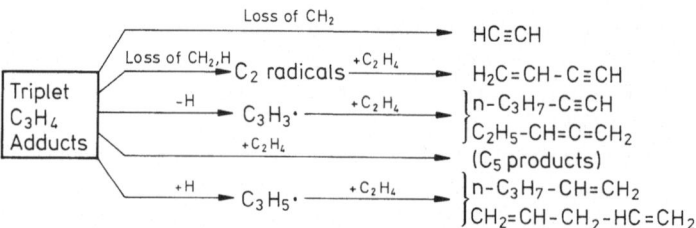

Addition of C (^{3}P) to ethylene. (Minor pathways are indicated by parenthesizing associated products.)

Fig. 3.1.9. Probable reaction modes of adducts formed by additions of C(^{1}D) and C(^{3}P) to ethylene. (Taken with permission from Ref. 3.1.94. Copyright (1964) American Chemical Society)

The reactions of recoil ^{11}C atom with ammonia give $^{11}CH_4$ and $^{11}CH_3NH_2$ as primary products [3.1.96]. Their yields are very sensitive to the radiation dose as shown in Fig. 3.1.10. Above 0.1 eV/molecule ($\sim 5 \times 10^7$ rad), product composition characteristic of the "saturation" region for this system is observed. Reduction of $^{11}CH_3NH_2$ to $^{11}CH_4$ cannot be brought about by the direct radiolysis of the former but the reduction should involve as the first step either interaction with charged or excited ammonia molecules resulting from the radiolysis of the total system.

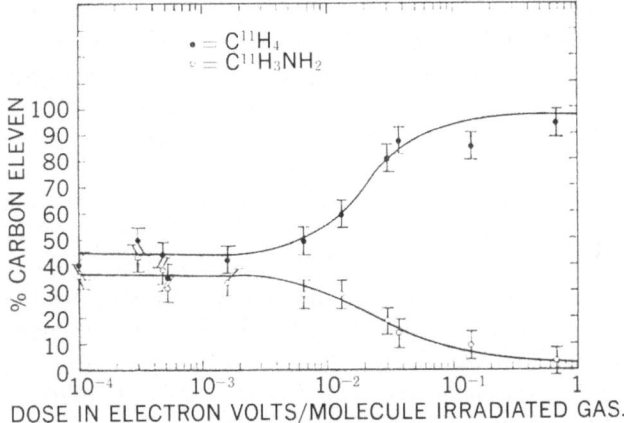

Fig. 3.1.10. Methane-^{11}C—methylamine-^{11}C yields as a function of dose to the system. (Reprinted with permission from Ref. 3.1.96. Copyright (1962) American Chemical Society)

In mixtures of hydrogen and ethylene [3.1.97], both methyne and methylene are formed. The former is inserted into the C—H bond of ethylene, forming excited allyl radical, which further reacts with another ethylene molecule to give 1-pentene-^{11}C. On the other hand, methylene-^{11}C is not inserted into C—H bond and can be readily scavenged by oxygen, indicating that it is in the triplet state [3.1.98].

3.1.9 Other Multivalent Elements

Silicon

Recoil silicon atoms can usually be generated by ^{30}Si(n, γ)^{31}Si, or ^{31}P(p, n)^{31}Si reaction. When a sufficient amount of phosphine is used as precursor of the recoiling silicon atoms, the mechanism for the product-forming steps involves silylene, ^{31}SiH$_2$, as a principal reactive intermediate which is inserted into Si—H bond, giving rise to the observed product. The possibility of hydrogen collection by the silicon atom will proceed as far as ^{31}SiH$_3$ seems to be quite small. Thus, in the reactions [3.1.99] with silane and its homologues under a variety of experimental conditions, the major component of the volatile product activity is always the next higher homologue of the starting material (60—95%), accounted for by assuming ^{31}SiH$_2$ insertion. In a 1:1 phosphine-silane mixture [3.1.100], the absolute yields of all products decrease by about 45% in the presence of 1 to 5% NO as scavenger, whereas the relative yields remain unaffected: about 3.6 as the ^{31}Si-disilane/^{31}Si-silane ratio. All product yields tend to become zero at 100% neon moderation, except for ^{31}Si-disilane and ^{31}S-trisilane which tend to approach finite values in the limit of 100% rare gas moderation.

As for the spin state of ^{31}SiH$_2$, a semiempirical calculation predicts [3.1.100] that the ^{1}A$_1$ singlet is the ground state of silylene while the ^{3}B$_1$ triplet state is much higher in energy (46 kcal/mol): the ^{1}A$_1$ state corresponds to the bond angle of 95°, and the first excited triplet state, to bond angle of 138°. For substituted silylenes [3.1.101] the ground state of dimethylsilylene is a singlet.

Tang et al. [3.1.102] have allowed $^{31}SiH_2$ to react with butadiene in a phosphine-butadiene (3:1) mixture. 3-silacyclopentene-^{31}Si (SCP) is formed normally corresponding to at least 50% of the total observed volatile activity: other products obtained include small amount of $^{31}SiH_4$ and varying amount of others. A typical scavenger curve shows that the specific activity of SCP decreases rapidly in the 0—4% NO region and remains constant thereafter: the plateau value attained is 20% of the original unscavenged value. Since NO removes effectively species with unpaired electrons, this observation may suggest that the reacting silylene is 80% in triplet and 20% in singlet. The apparent 1,4-addition of singlet $^{31}SiH_2$ to butadiene may be in fact 1,2-addition yielding unstable vinylsilacyclopropane followed by isomerization to SCP. The isomerization is likely to proceed fast, and therefore is unaffected by the presence of a scavenger. In order to ascertain the gound state of a divalent species, an efficient moderator was added to the system, and the results with neon as moderator has proved that the ground electronic state of silylene is a singlet.

Fig. 3.1.11. Product yields from reactions of recoil silicon atoms in phosphine-butadiene mixtures. Total pressure, 1000 Torr. □, $^{31}SiC_4H_8$ (3-silacyclopentene); ○, $^{31}SiC_4H_6$. (Reprinted with permission from Ref. 3.1.103. Copyright (1978) American Chemical Society)

Gaspar et al. [3.1.103] have further examined the reactions of ^{31}Si in various mixtures of phosphine and butadiene, and have observed that the yields of two major products, SCP and another product presumably 2,4-silacyclopentadiene-^{31}Si (SCD), depend upon the mole fraction of butadiene as shown in Fig. 3.1.11: the former is also obtained in 46% yield from the reaction of thermally generated SiH_2 with butadiene, while the latter is not. The possible mechanism leading to the formation of the latter will be:

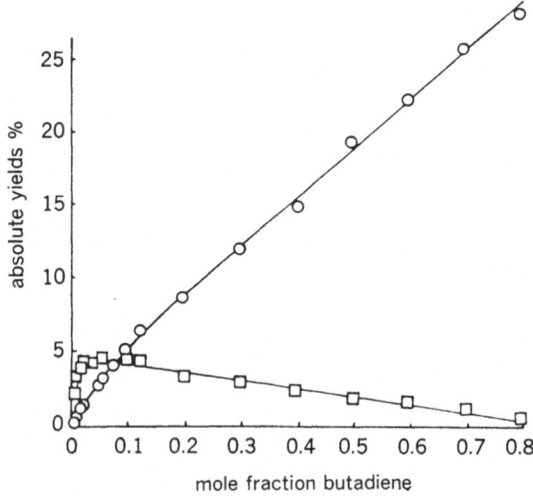

These results may raise a doubt as to the participation of ground state singlet $^{31}SiH_2$ as reactive intermediate although they do not absolutely exclude its possibility. The observed variation of product yields in Fig. 3.1.11 has revealed a competition between phosphine and butadiene for recoiling ^{31}Si atoms, if butadiene reacts to form SCD while the reaction with phosphine forms $^{31}SiH_2$, which then reacts with butadiene to give SCP. These results, together with those obtained with phosphine-silane-butadiene mixtures, will call for the following reaction processes:

$$^{31}Si + PH_3 \rightarrow \rightarrow {}^{31}SiH_2, \tag{3.1.75}$$

$$^{31}Si + SiH_4 \rightarrow \rightarrow {}^{31}SiH_2, \tag{3.1.76}$$

$$^{31}Si + SiH_4 \rightarrow H^{31}SiSiH_3, \tag{3.1.77}$$

$$^{31}Si + C_4H_6 \rightarrow \rightarrow {}^{31}SiC_4H_6(SCD), \tag{3.1.78}$$

$$^{31}SiH_2 + C_4H_6 \rightarrow {}^{31}SiC_4H_8(SCP), \tag{3.1.79}$$

$$H^{31}SiSiH_3 + SiH_4 \rightarrow SiH_3{}^{31}SiH_2SiH_3. \tag{3.1.80}$$

Sulfur

Recoil sulfur can be produced by $^{34}S(n, \gamma)^{35}S$, or $^{35}Cl(n, p)^{35}S$ reaction. In the gaseous mixtures of methane and hydrochloric acid [3.1.104], $H_2{}^{35}S$ and $CH_3{}^{35}SH$ are the two main products, whose yields are about 60% and 3%, respectively. The former yield is quite sensitive to NO as scavenger, indicating that the reactions involving the thermal radical mechanism are dominant for the $H_2{}^{35}S$ formation. The reactions responsible for the production of $CH_3{}^{35}SH$ appear to be fairly complicated. The $CH_3{}^{35}SH$ yield is somewhat increased by the addition of argon.

Rowland et al. [3.1.105] examined the reactions of ^{35}S in gaseous OCS-propane systems, in comparison with those of ^{35}S produced by the photolysis of $OC^{35}S$. The experiments in the photolysis are consistent with the explanation that both $S(^1D)$ and $S(^3P)$ atoms are present, and that no mercaptane formation is attributable to $S(^3P)$, which instead is efficiently removed by reaction with OCS. In the absence of OCS, the $S(^3P)$ atoms react with propane to form almost exclusively $(CH_3)_2CH^{35}SH$. The features obtained in the recoil experiments are the nonreproducibility of the yields and the always negligible yield of $n-C_3H_7{}^{35}SH$: the observed main products are $^{35}SO_2$ and $OC^{35}S$: both formed by reaction with trace amounts of oxygen and CO. Various amounts of $i-C_3H_7{}^{35}SH$ (0.1—10%) are also formed by wall reaction and/or radiation-induced reactions. Since the reaction of $^{35}S(^1D)$ with propane leads to the formation in good yield of $n-C_3H_7{}^{35}SH$, the absence of the latter in the OSC-propane systems convincingly demonstrates that negligible fraction of ^{35}S recoil atoms reacts as ^1D atoms in the chemical reaction range. Thus, it is suggested that the recoil ^{35}S atoms exist mostly in the ground electronic state, ^3P, and are non-reactive whilst hot in the gaseous phase. It may become reactive when adsorbed on the walls of reaction ampoules.

The high reactivity of recoil ^{35}S towards oxygen has also been observed in a variety of gaseous mixtures, H_2S, SO_2, SF_6, CH_3SH, and thiophene [3.1.106]. Only traces of oxygen in these gases result in the incorporation of quite large percentages of ^{35}S as $^{35}SO_2$. With H_2S, about 85% of the sulphur is found as

$H_2{}^{35}S$, and this percentage is not affected by the addition of argon but reduced to 3% in the presence of oxygen, $H_2{}^{35}S$ is also a major product from CH_3SH and thiophene. In the absence of oxygen, ca. 90 and 70% of ^{35}S formed exist in the gaseous fraction.

In the reaction with CH_3Cl [3.1.104], $H_2{}^{35}S$ ($\sim 2\%$), $CH_3{}^{35}SH$ (5%), and $(CH_3)_2{}^{35}S$ are formed, and the last compound shows a pressure dependence, increasing from 13.4% at 200 mmHg to 24.5% at 500 mmHg. With C_2H_5Cl, a wide variety of products are obtained, among which $(C_2H_5)_2{}^{35}S$ is the major one (16% at 500 mmHg).

The complexing reaction processes of ^{35}S are still largely unclear; the probable mechanism would involve at least two steps, corresponding to the formation of two chemical bonds. This might be written,

$$^{35}S + [X-Y] \rightarrow [X-{}^{35}S-Y] \rightarrow \text{products} \qquad (3.1.81)$$

where X and Y stand for hydrogen, alkyl group or any other species.

Phosphorus

Very little work has been made on the hot atom chemistry of phosphorus in the gas phase. In the reaction of recoil ^{32}P from the $^{31}P(n, \gamma)^{32}P$ reaction with PF_3 [3.1.107], the only products observed was $^{32}PF_3$ with a yield of 38%. In mixtures of PF_3 and H_2, $^{32}PH_3$ was observed as a product. The yield increased from 7% to 20% as the H_2-concentration was changed from 50 to 71%. $^{32}PF_3$ yield is decreased concomitantly from 17% to less than 1%. This clearly indicates the reactivity of ^{32}P toward H_2 to give $^{32}PH_3$. Various additives can serve as scavengers in the reaction of ^{32}P with PF_3 as shown in Table 3.1.16. In the PF_3-PH_3 systems, the sharp increase in $^{32}PH_3$ occured at the expense of $^{32}PF_3$; no activities corresponding to the products $^{32}PFH_2$ or $^{32}PHF_2$ were observed. The sum of the yields of both products as a function of the amount of PH_3 lies along the line connecting the yields in two extreme reaction systems. This indicates that PH_3 scavenges all ^{32}P which would otherwise react with PF_3 and that the scavenged product is $^{32}PH_3$. The moderator curve for $^{32}PF_3$ in the presence of 7% PH_3, shows that the $^{32}PF_3$ is formed by thermal reactions, whereas 60% of $^{32}PH_3$ is formed by hot reactions and 40% by thermal reactions.

Table 3.1.16. $^{32}PF_3$ yields from PF_3 (taken from Ref. 3.1.107)

Added gas (vol.-%)	$^{32}PF_3$, %	Added gas (vol.-%)	$^{32}PF_3$, %
None	38	PH_3 (7) + H_2S (68)	0 ± 0.1
H_2 (1)	42	C_6H_{12} (8)	0 ± 0.1
H_2 (50)	17	C_6H_{12} (10.5)	0 ± 0.1
H_2 (71)	0 ± 0.2	C=C-C=C (7.75)	0 ± 0.1
H_2S (3)	9	C=C-C=C (24)	0 ± 0.1
H_2S (76)	0 ± 0.5	CF_4 (8)	19
PH_3 (3)	4	CF_4 (20)	17
PH_3 (50)	0.4 ± 0.2	PH_3 (7) + C=C (5)	1.6
PH_3 (76)	1.5 ± 0.2	PH_3 (7) + C=C (40)	0.6
PH_3 (7) $\pm$ H_2S (44)	0 ± 0.1		

Table 3.1.17. Yields of ^{75}Ge-labeled products from reactions of ^{75}Ge atoms^a (taken from Ref. 3.1.109; Copyright 1969, American Chemical Society)

Expt.	Substrate	^{75}GeH$_4$^{b}	H$_3$^{75}GeGeH$_3$^{b}	H$_3$^{75}GeSiH$_3$^{b}	$\dfrac{H_3^{75}GeGeH_3}{^{75}GeH_4}$	$\dfrac{H_3^{75}GeGeH_3}{H_3^{75}GeSiH_3}$
A	GeH$_4$	3,900 ± 600	17,300 ± 2,800	—	4.5 ± 1.0	—
A	1:1 GeH$_4$–SiH$_4$	1,500 ± 200	7,600 ± 1,200	1,900 ± 300	5.2 ± 1.2	4.0 ± 0.9
B	GeH$_4$	2,200 ± 400	8,300 ± 1,300	—	3.8 ± 0.9	—
B	GeH$_4$ + 10% NO	3,800 ± 600	5,400 ± 900	—	1.4 ± 0.3	—

^{a} All samples at total pressure 2.2 ± 0.1 atm and, within an experiment, irradiated simultaneously at equal neutron flux.
^{b} Yields given as counts above background extrapolated to a common time for each experiment to correct for decay.

Tang et al. [3.1.110] have further studied the moderator effects on these two types of ^{32}P-abstraction; the ^{32}PF$_3$ yield from PF$_3$ was not diminished by moderation, while the ^{32}PH$_3$ yield from PH$_3$ was reduced to as low as 27% of the unmoderated value, in the presence of 98% Ne. The following conclusion has been obtained: the H-abstraction from PH$_3$ has a reaction cross section curve covering wide energy range with rather low threshold. The corresponding F-abstraction appears to involve thermal ^{32}P reactions with much narrower energy range for the reaction cross section curve.

The interpretation of these results is not based on one-step formation process, while it consists in step wise abstraction of H or F atoms from the reactant molecules, thus being initiated by

$$^{32}P + PF_3 \rightarrow {}^{32}PF + PF_2, \tag{3.1.82}$$

$$^{32}P + PH_3 \rightarrow {}^{32}PH + PH_2. \tag{3.1.83}$$

Although these reactions are endothermic, ΔH should be small.

Trimethylphosphine as reactant [3.1.108] gives small yields of mono- (3%), di- (0.4%) and tri- (2.4%) methyl phosphines. The remainder of the activity is found on the walls of ampoules. The activity in trimethylphosphine is decreased with lowered pressure, thus demonstrating complete bond rupture following a neutron capture event.

Germanium

Striking similarity between germanium and silicon atoms has been observed in analogous reaction systems, although the total yield of volatile products was less in the former. When germanium atoms were produced from gaseous germane by recoil from the nuclear transformation ^{76}Ge(n, 2n)^{75}Ge [3.1.109], the product ratio H$_3$^{75}GeGeH$_3$/^{75}GeH$_4$ = 5.0 $\pm$ 0.8 (Table 3.1.17) is identical within experimental error with the product ratio H$_3$^{31}SiSiH$_3$/^{31}SiH$_4$ = 4.8 $\pm$ 0.9 obtained in the silane-rich phosphine-silane mixtures irradiated with fast neutrons [3.1.111]. The absolute yield of volatile products from the germanium system (20 $\pm$ 10%) is lower than that obtainable in the silicon system (> 80%). The presence of silane in the reaction mixture may neither change the total yield of volatile products per unit of precursor, nor the ratio of H$_3$^{75}GeGeH$_3$/^{75}GeH$_4$. This suggests that ^{75}Ge can react with silane giving silygermane as well as reacting with germane. The greater affinity for germane over silane might be related with the relative bond strength of Ge$-$H and Si$-$H. With these limited experiments, the germylene radical, :^{75}GeH$_2$, has been suggested as an important reacting species.

$$:{}^{75}GeH_2 + GeH_4 \rightarrow H_3{}^{75}GeGeH_3, \tag{3.1.84}$$

$$:{}^{75}GeH_2 + GeH_4 \rightarrow {}^{75}GeH_4 + :GeH_2, \tag{3.1.85}$$

$$:{}^{75}GeH_2 + SiH_4 \rightarrow H_3{}^{75}GeSiH_3. \tag{3.1.86}$$

References

[3.1.1]   Wolfgang, R.: Prog. Reaction Kinetics *3*, 97 (1964)
[3.1.2]   Urch, D. S.: MTR. Int. Rev. Sci., Inorg. Chem. Ser. Two, *8*, 49 (1975)
[3.1.3]   Urch, D. S., Welch, M. J.: Trans. Faraday Soc. *64*, 1547 (1968)
[3.1.4]   Malcolme-Lawes, D. J., Urch, D. S.: J. Chem. Soc., Faraday Trans. II *68*, 967
          (1972)
[3.1.5]   Carlson, A., Freedman, A., Press, G. A., Malcolme-Lawes, D. J.: Radiochim.
          Acta *18*, 167 (1972)
[3.1.6]   Baer, M., Amiel, S.: J. Chem. Phys. *53*, 407 (1970)
[3.1.7]   Alfassi, Z. B., Amiel, S.: ibid. *57*, 5085 (1972)
[3.1.8]   Keizer, J.: J. Chem. Phys. *56*, 5985 (1972)
[3.1.9]   Keizer, J.: ibid. *58*, 4524 (1973)
[3.1.10] Porter, R. N.: ibid. *45*, 2284 (1966)
[3.1.11] Huang, K.: *Statistical Mechanics*. New York: Wiley 1963, Ch. 6
[3.1.12] Grant, E. R., Feng, D.-F., Keizer, J., Knierim, K. D., Root, J. W.: in *Fluorine-
          Containing Free Radicals: Kinetics and Dynamics of Reactions* J. W. Root (ed.),
          ACS Symposium Series No. 66, American Chemical Society, Washington, D. C.,
          1978, p. 340
[3.1.13] Kushner, R., Rowland, F. S.: J. Amer. Chem. Soc. *91*, 1539 (1969)
[3.1.14] Rowland, F. S.: in *Molecular Beam Reaction Kinetics* Ch. Schlier (ed.). New York:
          Academic Press 1970
[3.1.15] Wolfgang, R.: Ann. Rev. Phys. Chem. *16*, 15 (1965)
[3.1.16] Urch, D. S.: MTP Int. Rev. Sci., Inorg. Chem. Ser., One *8*, 149 (1972)
[3.1.17] Rowland, F. S.: MTP Int. Rev., Phys. Chem. Ser., One *9*, 109 (1972)
[3.1.18] Tang, Y. N.: in *Isotopes in Organic Chemistry* E. Buncel, C. C. Lee (eds.). Elsevier
          1979, p. 85
[3.1.19] Lee, E. K. C., Rowland, F. S.: J. Amer. Chem. Soc. *85*, 897 (1963)
[3.1.20] Robinson, P. J., Holbrook, K. A.: *Unimolecular Reactions*. Wiley-Interscience
          1971, p. 64
[3.1.21] Wolf, A. P., Pettijohn, R. P., To, K. C., Rack, E. P.: J. Phys. Chem. *83*, 1237
          (1979)
[3.1.22] Lee, E. K. C., Miller, G., Rowland, F. S.: J. Am. Chem. Soc. *87*, 190 (1965)
[3.1.23] Witkin, J., Wolfgang, R.: J. Phys. Chem. *72*, 2631 (1968)
[3.1.24] Tominaga, T., Hosaka, A., Rowland, F. S.: ibid. *73*, 465 (1969)
[3.1.25] Smith, W. S., Tang, Y. N.: J. Chem. Phys. *78*, 2186 (1974)
[3.1.26] Lee, E. K. C., Rowland, F. S.: J. Amer. Chem. Soc. *85*, 897 (1963)
[3.1.27] Kushner, R., Rowland, F. S.: J. Phys. Chem. *76*, 190 (1972)
[3.1.28] Ferro, L. J., Spicer, L. D.: J. Chem. Phys. *69*, 4335 (1978)
[3.1.29] Palino, G. P., Rowland, F. S.: J. Phys. Chem. *75*, 1299 (1971)
[3.1.30] Valencich, T., Bunker, D. L.: Chem. Phys. Lett. *20*, 50 (1973)
[3.1.31] Tang, Y. N., Lee, E. K. C., Tachikawa, E., Rowland, F. S.: J. Phys. Chem. *74*,
          675 (1970)
[3.1.32] Chou, C. C., Wilkey, D. D., Rowland, F. S.: Chem. Phys. Lett. *20*, 53 (1973)
[3.1.33] Chou, C. C., Rowland, F. S.: J. Phys. Chem. *75*, 1283 (1971)
[3.1.34] Compton, L. E., Veberly, G. D., Martin, R. M.: ibid. *78*, 56 (1974)
[3.1.35] Beverly, G. D., Martin, M.: ibid. *80*, 2063 (1976)
[3.1.36] Kunty, P. J., Nemeth, E. M., Polany, J. C., Wong, W. H.: J. Chem. Phys. *52*,
          4654 (1970)
[3.1.37] Urch, D. S., Wolfgang, R.: J. Amer. Chem. Soc. *83*, 2982 (1961)
[3.1.38] Root, J. W., Breckenridge, W., Rowland, F. S.: J. Chem. Phys. *43*, 3964 (1963)
[3.1.39] Tachikawa, E., Rowland, F. S.: J. Amer. Chem. Soc. *91*, 559 (1969)
[3.1.40] Tachikawa, E., Tang, Y. N., Rowland, F. S.: ibid. *90*, 3584 (1968)
[3.1.41] Tominaga, T., Rowland, F. S.: J. Phys. Chem. *72*, 1399 (1968)
[3.1.42] Chou, C. C., Rowland, F. S.: J. Phys. Chem. *75*, 1283 (1971)

[3.1.43] Chou, C. C., Smail, T., Rowland, F. S.: J. Amer. Chem. Soc. *91*, 3104 (1969)
[3.1.44] Urch, D. S., Welch, M. J., Arthy, R. J.: Trans. Faraday Soc. *66*, 1642 (1970)
[3.1.45] Smail, T., Arezzo, B., Rowland, F. S.: J. Phys. Chem. *76*, 187 (1972)
[3.1.46] Celebourne, N., Wolfgang, R.: J. Chem. Phys. *38*, 2782 (1963)
[3.1.47] Iyer, R. S.: Ph.D. Thesis, University of California, Irvine (1973)
[3.1.48] Smail, T., Iyer, R. S., Rowland, F. S.: J. Amer. Chem. Soc. *94*, 1041 (1972)
[3.1.49] Parks, N. J., Krohn, K. A., Root, J. W.: J. Chem. Phys. *55*, 2690 (1971)
[3.1.50] Williams, R., Rowland, F. S.: J. Phys. Chem. *77*, 301 (1973)
[3.1.51] Tang, Y. N., Smail, T., Rowland, F. S.: J. Amer. Chem. Soc. *91*, 2130 (1969)
[3.1.52] Smail, T., Miller, G. E., Rowland, F. S.: J. Phys. Chem. *74*, 3464 (1970)
[3.1.53] McKnight, C. F., Root, J. W.: J. Phys. Chem. *73*, 4330 (1969)
[3.1.54] Celebourne, N., Todd, J. F. J., Wolfgang, R. L.: Chem. Effects Nucl. Transform.,
         Vienna 1, 149 (1965), Vienna, IAEA
[3.1.55] Spicer, L. D., Sinda, A.: Radiochim. Acta *18*, 16 (1972)
[3.1.56] Smail, T., Rowland, F. S.: J. Phys., Chem. *74*, 1866 (1970)
[3.1.57] McKnight, C. F., Parks, N. J., Root, J. W.: J. Phys. Chem. *74*, 217 (1970)
[3.1.58] Krohn, K. A., Parks, N. J., Root, J. W.: J. Chem. Phys. *55*, 5771 (1971)
[3.1.59] Krohn, K. A., Parks, N. J., Root, J. W.: J. Chem. Phys. *55*, 5785 (1971)
[3.1.60] Root, J. W.: J. Phys. Chem. *73*, 3174 (1969)
[3.1.61] Palino, G. F., Rowland, F. S.: Radiochim. Acta. *15*, 57 (1971)
[3.1.62] Bunker, D. L.: J. Chem. Phys. *57*, 332 (1972)
[3.1.63] Smail, T., Iyer, R. S., Rowland, F. S.: J. Phys. Chem. *75*, 1324 (1971)
[3.1.64] Wai, C. M., Rowland, F. S.: J. Phys. Chem. *71*, 2752 (1967)
[3.1.65] Alfassi, Z. B., Amiel, S.: Radiochim. Acta *15*, 201 (1971)
[3.1.66] Spicer, L. D., Wolfgang, R.: J. Amer. Chem. Soc. *90*, 2426 (1968)
[3.1.67] Spicer, L. D.: ibid. *95*, 51 (1973)
[3.1.68] Spicer, L. D., Wolfgang, R.: J. Chem. Phys. *50*, 3466 (1969).
[3.1.69] Tang, Y. N., Smith, W. S., Williams, J. L., Lowery, K., Rowland, F. S.: J.
         Phys. Chem. *75*, 440 (1971)
[3.1.70] Stevens, D. J., Spicer, L. D.: J. Phys. Chem. *82*, 627 (1978)
[3.1.71] Wolf, A. P., Schuler, P., Pettijohn, R. P., To, K. C., Rack, E. P.: J. Phys. Chem.
         83, 1237 (1979)
[3.1.72] Helton, R. W., Yoong, M., Rack, E. P.: J. Phys. Chem. *75*, 2072 (1971)
[3.1.73] Saeki, M., Tachikawa, E.: Bull. Chem. Soc. Jpn. *46*, 839 (1973)
[3.1.74] Tachikawa, E., Yanai, K.: Radiochim. Acta *17*, 138 (1972)
[3.1.75] Numakura, K., Tachikawa, E.: Bull. Chem. Soc. Jpn. *46*, 346 (1973)
[3.1.76] Tachikawa, E., Numakura, K.: ibid. *47*, 2749 (1974)
[3.1.77] Kondo, K., Yagi, M.: ibid. *51*, 1284 (1978)
[3.1.78] Saeki, M., Tachikawa, E.: J. Chem. Soc., Faraday Trans. I *71*, 2121 (1975)
[3.1.79] Rack, E. P., Gordus, A. A.: J. Chem. Phys. *34*, 1855 (1961)
[3.1.80] Loberg, M. D., Welch, M. J.: J. Amer. Chem. Soc. *95*, 1073 (1973)
[3.1.81] Saeki, M., Tachikawa, E.: Bull. Chem. Soc. Jpn. *50*, 1762 (1977)
[3.1.82] Nicholas, J. B., Yoong, M., Rack, E. P.: Radiochim. Acta *19*, 124 (1973)
[3.1.83] Lararde, P. G., Abbe, J. C., Paulus, J. M.: Radiochim. Acta *17*, 96 (1972)
[3.1.84] Pettijohn, R. R., Rack, E. P.: J. Phys. Chem. *76*, 3342 (1972)
[3.1.85] Yoong, M., Pao, Y. C., Rack, E. P.: ibid. *76*, 2685 (1972)
[3.1.86] Berg, M. E., Loventhal, A., Adelman, D. J., Grauer, W. M., Rack, E. P.: ibid.
         80, 837 (1977)
[3.1.87] Ache, H. J., Christman, D. R., Wolf, A. P.: Radiochim. Acta *12*, 121 (1969)
[3.1.88] Lambrecht, R. M., Furukawa, N., Wolf, A. P.: J. Phys. Chem. *74* 4605 (1970)
[3.1.89] Welch, M. J., Wolf, A. P.: J. Amer. Chem. Soc. *91*, 6584 (1969)
[3.1.90] Stöcklin, G., Wolf, A. P.: Chem. Effects Nucl. Transform Vol. 1, IAEA, 1965,
         p. 121
[3.1.91] Taylor, K. K., Ache, H. J., Wolf, A. P.: J. Amer. Chem. Soc. *97*, 5970 (1975)
[3.1.92] Taylor, K. K., Ache, H. J., Wolf, A. P.: J. Phys. Chem. *82*, 2385 (1978)
[3.1.93] Marchall, M., MacKay, C., Wolfgang, R.: J. Amer. Chem. Soc. *86*, 4741 (1964)
[3.1.94] Dubrin, J., MacKay, C., Wolfgang, R.: ibid. *86*, 4747 (1964)

[3.1.95] Dubrin, J., MacKay, C., Pandow, M. L., Wolfgang, R.: J. Inorg. Nucl. Chem. 26, 2113 (1964)

[3.1.96] Cacace, F., Wolf, A. P.: J. Amer. Chem. Soc. 84, 3202 (1962)

[3.1.97] Nicholas, J., MacKay, C., Wolfgang, R.: ibid. 88, 1065 (1966)

[3.1.98] Rose, T. L.: J. Phys. Chem. 76, 1389 (1972)

[3.1.99] Gaspar, P. P., Markush, P., Holten, J. D., III, Frost, J. J.: J. Phys. Chem. 76, 1352 (1972)

[3.1.100] Jordan, P. C.: J. Chem. Phys. 44, 3400 (1966)

[3.1.101] Skell, P. S., Goldstein, E. J.: J. Amer. Chem. Soc. 86, 1442 (1964)

[3.1.102] Zeck, O. F., Su, Y. Y., Gennaro, G. P., Tang, Y. N.: ibid. 96, 5967 (1974)

[3.1.103] Hwang, R. J., Gaspar, P. P.: ibid. 100, 6626 (1978)

[3.1.104] Panek, K., Mudra, K.: Symp. Chem. Effects Nucl. Transform. Vienna, Vol. 1, (1965), p. 195, Vienna, IAEA

[3.1.105] Church, L. B., Rowland, F. S.: Radiochim. Acta 16, 55 (1971)

[3.1.106] Hyder, M. L., Markowitz, S. S.: J. Inorg. Nucl. Chem. 26, 257 (1964)

[3.1.107] Stewart, G. W., Hower, C. O.: J. Inorg. Nucl. Chem. 34, 39 (1972)

[3.1.108] Halmann, M.: Symp. Chem. Effects Nucl. Transform., Prague, Vol. 1, (1961), p. 195, IAEA

[3.1.109] Gaspar, P. P., Levy, C. A., Frost, J. J., Bock, S. A.: J. Amer. Chem. Soc. 91, 1573 (1969)

[3.1.110] Zeck, O. F., Ferriere, R. A., Copp, C. A., Gennaro, G. P., Tang, Y. N.: J. Inorg. Nucl. Chem. 41, 785 (1979)

[3.1.111] Gaspar, P. P., Bock, S. A., Eckelman, W. C.: J. Amer. Chem. Soc. 90, 6914 (1968)

3.2 Liquid Phase Hot Atom Reactions

3.2.1 Organic Condensed Phase

The mechanisms of hot-atom reactions in the gas phase are more or less common and relatively straightforward. Condensed phase hot atom phenomena, however, are less understood. While primary reactions of hot atoms are normally relatively phase-independent, actual yield spectra of hot atom processes are often drastically different in gas and condensed phase media. In the condensed phase, the collision times become longer: typical collision times are about a factor of 100 greater than in the gas phase at normal pressures. The molecules tightly packed are kept together and the time of interaction is prolonged. In addition, the solvation and efficient energy transfer may promote reaction processes which would not occur in the gas phase. Accordingly, the following factors become important in the condensed phase:

a) Collisional stabilization

Unimolecular decomposition of excited primary products competes with collisional stabilization. The high collision density in the condensed phase shows decompoition persisting.

b) Cage effect [3.2.1]

Hot atoms with excess energies rupture bonds during the energy dissipation process. In the gas phase, the products from such processes will be separated instantly from each other, whereas in the condensed phase, they may be confined together in the same solvent cage. The hot atoms may eventually undergo a secondary recombination process to yield the identical products with those resulted from the direct processes.

Density Effect. Figure 3.2.1 shows the results of the pioneering work by Richardson and Wolfgang [3.2.2] on the dependency of the yields of $CH_3{}^{18}F$ and $CH_2{}^{18}FF$ on the density of reaction system from gas to close-packed liquid, or solid. Two products arise from ^{18}F-for-F and ^{18}F-for-H reactions in the I_2-CH_3F systems. The large density change could only be achieved by changing the reaction temperature during the irradiation. However, the temperature change will not affect the

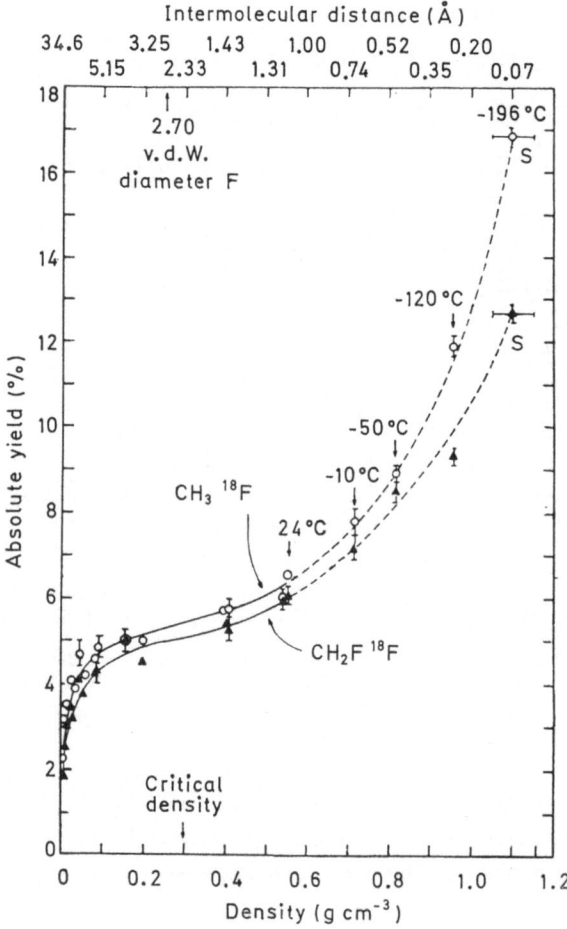

Fig. 3.2.1. Absolute yields of single displacement products as a function of density in the CH_3F-I_2 system. $CH_3{}^{18}F$, ○; $CH_2F{}^{18}F$, ▲. Temperature at 55°C unless otherwise specified. Iodine either at vapor pressure at 55°C or at saturated solution concentration at the specified temperature. *S* indicates sample was solid, otherwise it was fluid. (Reprinted with permission from Ref. 3.2.2. Copyright (1970) American Chemical Society)

product yields within the experimental error [3.2.3] and the results in the figure are essentially due to the density variation. Both products show a sharp rise in yield from the lowest gas density, and level off at about 0.2 g cm^{-3} (50—100 atm). Such behavior is characteristic of increased collisional stabilization of primary products left with high internal energy which otherwise may decompose unimolecularly. Of even greater magnitude is the second rise in yield at higher densities. This second rise is due to the cage effect, and indicates that caging becomes important only when the mean intermolecular distance is reduced to about half the diameter of the fluorine atom. This will be conceivable, since there will always be some larger interstices in the cage wall through which the energetic atom can squeeze.

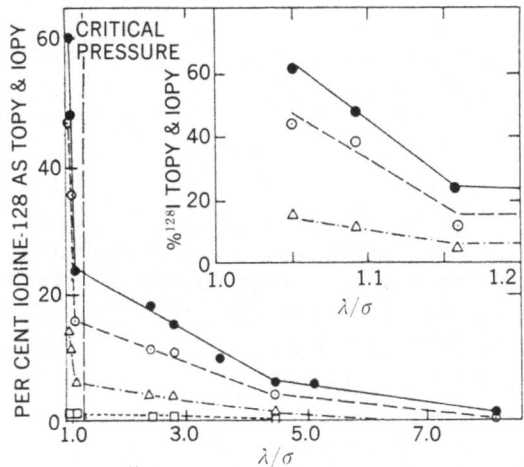

Fig. 3.2.2. Total and individual organic product yields vs. λ/σ. — • —, total organic product yield; -- ○ --, $C_2H_5{}^{128}I$; ···△···, $CH_3{}^{128}I$; ···□···, $C_2H_3{}^{128}I$. Insert shows expansion of region $\lambda/\sigma < 1.2$ for TOPY, ethyl iodide and methyl iodide. (Reprinted with permission from Ref. 3.2.4. Copyright (1977) American Chemical Society)

In ethane atmosphere, more than 50% of the (n, γ)-activated ^{128}I is ionic. Figure 3.2.2 [3.2.14] depicts the total (TOPY) and individual (IOPY) organic product yields of the ^{128}I — ethane system as functions of λ/σ: the intermolecular distance, λ, is calculated by

$$\lambda = (\rho N/M)^{-1/3} \qquad (3.2.1)$$

where ρ is the density in g · cm^{-3}, N the Avogadro's number, and M the molecular weight in grams. The value of σ used is 4.54 A. Three regions are observed: (i) at low to intermediate density, a rapid increase in product yield as a result of increased stabilization of molecular reaction products; (ii) at intermediate density, an approximately linear region with rapid stabilization; and (iii) at higher density, a region with rapidly increasing product yield resulting from continued collisional stabilization of highly excited product moieties and occurence of enhancement reactions. In view of various possible ion-molecule reactions, it is believed that the formation of methyl iodide is primarily ascribed to direct substitution which exhibits "cage-radical" or "cage-ion" enhancement, while the vinyl- and ethyl iodides result from a complexation reaction, or enhancement by a "caged-complex".

Cage Effect. Although the rises in product yields are usually observed with increasing density in reactions of recoil halogen atoms, there are two types of caged recombinations: (i) geminate recombination—recombination of the energetic atom with an original radical formed in the nuclear event, and (ii) radical recombination—reassembly of the atom with organic fragments in the vicinity of the nuclear event. Assuming that the extent of geminate recombination decreases with increased dilution of the substrate, Rack et al. [3.2.3] examined the reactions of recoil ^{128}I with dilute aqueous mixtures of mono-iodotyrosine (MIT) and diiodotyrosine (DIT), where water was used as a reactive solvent. Thus, ^{128}I$^-$ is formed on subsequent ionization of H^{128}I resulting from hydrogen abstraction from a water molecule. When MIT and DIT are diluted from 10^{-2} mol% down to -10^{-4} mol% in the liquid phase, the ^{128}I-for-I reaction product yields approach zero, while the ^{128}I$^-$ yield increases to 100%. In frozen solutions, however, the substitution yields remain constant at $\sim 17\%$ over a 100-fold concentration range. In addition, for the MIT system, the ^{128}I$^-$ yield appears constant at $\sim 75\%$ even at the greatest dilution. Water molecules in ice are in hexagonal array through hydrogen bonding, and the DIT or MIT molecules are trapped in these empty spaces between the molecules. Under these circumstances, recoil ^{128}I remains close to the organic radical by virtue of the efficient caging ability of the ice lattice and undergoes a geminate recombination. Hence, the cage recombination may depend upon mass, viscosity and temperature, which affect the rate of out-of-cage diffusion. However, Stöcklin et al. [3.2.5] have suggested that it will not be necessary to expect substantial contribution of radical-radical combination processes in the reaction of recoil F and Cl atoms, since the thermalized atoms are readily removed by the exothermic hydrogen abstraction reaction,

$$\text{F (or Cl)} + \text{RH} \rightarrow \text{R}\cdot + \text{HF (or HCl)}. \tag{3.2.2}$$

The activation energy for the gas phase abstraction of a primary hydrogen by Cl atom is about 1 kcal/mol, and F and Cl atoms will have only a small chance for any type of radical combination reactions, which can hardly compete with fast hydrogen abstraction. Hence, for the reaction of ^{38}Cl with rac- and meso-2,3-dichlorobutane in the liquid phase, total organic yields remain unaffected by iodine scavenger over the entire concentration range up to saturation [3.2.6]. This does not verify the possibility of the presence of immediate cage combination reactions, but demonstrates the absence of thermal diffusive radical reactions and the self-scavenging properties of the system. Under these conditions, the stereochemical course of hot ^{38}Cl-for-Cl reaction is significantly influenced by changing the type and the concentration of solvent, as seen in Fig. 3.2.3. The magnitude of the effects is quite different for meso- and racemic compounds, and the decrease or increase in retention is more pronounced in the case of racemic form as parent. The most likely explanation for the solvent effect is the change in stereochemical influence caused by the change in rotational isomer concentration. Stöcklin et al. [3.2.6] have considered that such conformational effect provides evidence for one-step substitution, which can lead to both retention and inversion in the liquid phase. They [3.2.7] have further postulated that the hot substitution can proceed via two channels, i.e., direct replacement without the change in configuration, and formation of a longer lived complex with a lifetime long enough to allow inversion of the configuration. While the direct reaction should be predominant at lower pressures, the complex formation, if it occurs at all, is most likely to take place at higher pressures, especially in the liquid phase, because of the collisional stabilization and the prolonged collision time.

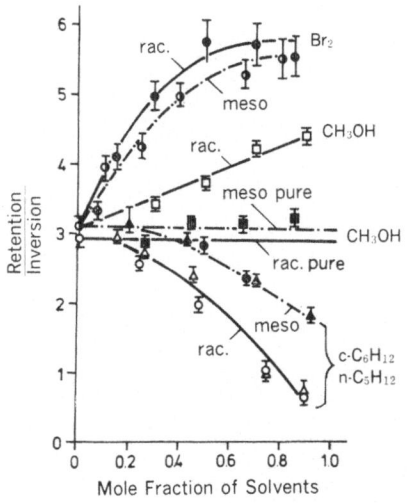

Fig. 3.2.3. Solvent effect on the stereochemical course (ratio retention/inversion) of ^{38}Cl-for-Cl substitution in rac- and meso-2,3-DCB. $\otimes$, rac-DCB-Br$_2$; $\square$, rac-DCB-CH$_3$OH; $\triangle$, rac-DCB-n-C$_5$H$_{12}$; $\bigcirc$, rac-DCB-c-C$_6$H$_{12}$; $\bullet$, meso-DCB-Br$_2$; $\blacksquare$, meso-DCB-CH$_3$OH; $\blacktriangle$, meso-DCB-n-C$_5$H$_{12}$; $\bullet$, meso-DCB-c-C$_6$H$_{12}$. (Reprinted with permission from Ref. 3.2.6. Copyright (1972) American Chemical Society)

Subsequently they have compared the pressure and phase dependency of the T-for-H and ^{38}Cl-for-Cl reaction yields in both rac- and meso-(CHClF)$_2$ [3.2.5]. As seen in Fig. 3.2.4, the phase effect is not important for recoil tritium because of its smaller size. The effect takes part in the ^{38}Cl-for-Cl reaction and is almost identical in both diastereomeric substrates. Although the absolute yield of the retained substitution product increases only by a factor of about 2.5 as compared with roughly 20 for the inverted product, over the range from the high-pressure gas (about 10^4 torr) to the liquid phase in both systems, the additional increase for the other products found in the liquid phase is slightly higher for the retained than for the inverted substitution product (namely, 2.3 as compared with 1.5 in the meso-, and 1.8 as compared with 1.3 in the rac-(CHClF)$_2$ system.) It is noted, however, that the yield for the retained form already starts to increase in the high-pressure range in the gas phase, and simply continues to increase further almost linearly with increasing density, whereas significant inversion is observed only in the liquid phase. The different dependence on density between the retained and inverted ^{38}Cl-for-Cl reaction yield is inconsistent with the caged combination, which should lead to racemization, and the 'caged complex' mechanism has been proposed. In this mechanism there is only one adduct (perhaps electronically unstable) formed from the hot atom and a substrate molecule, the lifetime of which is effectively prolonged by the surrounding molecules. The substitution processes leading both to retention

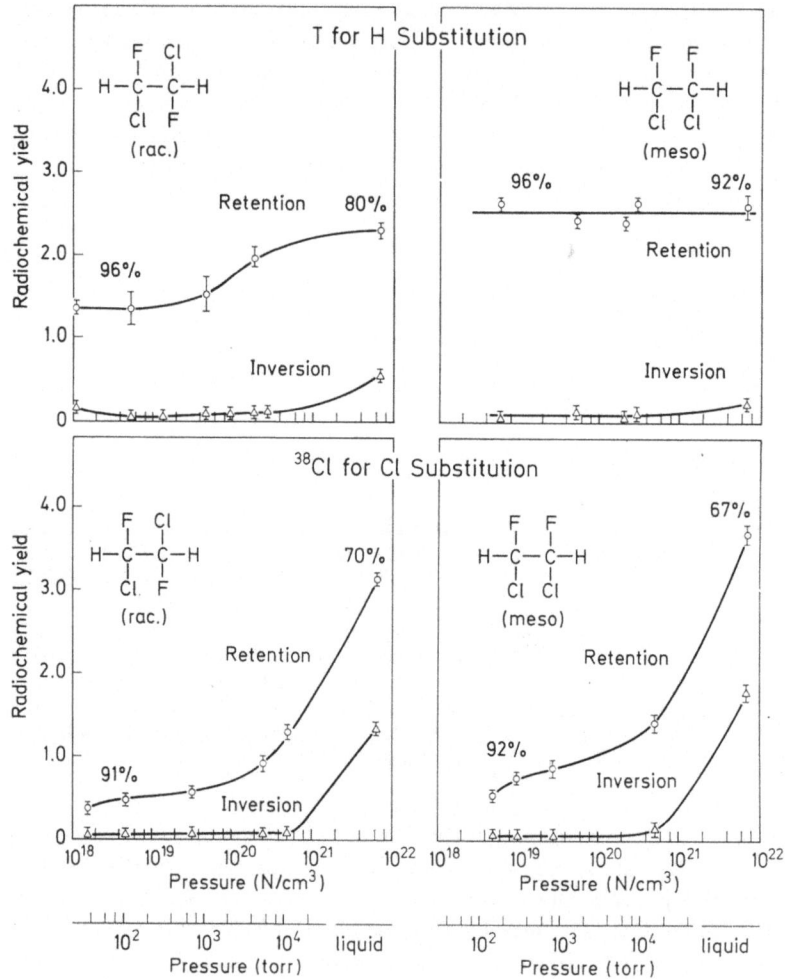

Fig. 3.2.4. Comparison of pressure and phase dependence of hot T-for-H and ^{38}Cl-for-Cl substitution yields in rac- and meso-(CHClF)$_2$. (Reprinted with permission from Ref. 3.2.5. Copyright (1974) American Chemical Society)

and inversion of configuration would occur concomitantly, but racemization via a planar radical will not be possible.

Later, Wolf et al. [3.2.8] have indicated that condensed phase stereochemical experiments with diastereomers are difficult to interpret due to the difference in the thermodynamic stabilities between the compounds in meso- and racemic modifications. The energies involved in recombination reactions can be different between the retained and inverted products. This difference may lead to a distorted explanation for the stereochemical course of the replacement reactions taking place. Enantiomeric molecules offer a means of studying replacement reactions in which both the initial and the final energy states of the reactions are thermodynamically identical. For the ^{36}Cl- and ^{34m}Cl-for-Cl reactions at the asymmetric carbon of 2(S)-(+) and 2(R)-(−)-chloropropionyl chloride in the condensed phase, the retention of optical configuration is 50%. A similar value has also been obtained with 2(S)-chloro-4-methylvaleryl chloride, in which the backside attack is sterically hindered significantly. Thus, the large increase in the retention would be caused by direct hot Cl-for-Cl replacement mechanism if involved. Based on these results, Wolf et al. [3.2.8] have concluded that racemization occurs through radical-radical recombination in the liquid, glassy and crystalline solid. The presence of both labeled isomers in equal amounts is suggestive of the fact that planarity of the radicals is established before recombination.

3.2.2 Inorganic Liquid Systems

We may now examine typical examples of inorganic liquid systems, i.e., solutions of metal complexes as well as organometallic compounds. While extensive work has been done on the recoil and subsequent reactions in inorganic solids, relatively few detailed studies have been contributed to the chemistry of recoil species in inorganic liquid systems.

The recoil reactions in liquid phase are generally complicated, as compared with those in gas phase, because of the interactions of the solvent with the recoil species (e.g., energy transfer mechanism, solvation, and radical formation). However, such solvent effects may provide clues to elucidation of the mechanisms of recoil reactions; unlike with the solid phase reactions, we can elucidate the mechanisms of liquid phase reactions by studying the effects of various solvents (nature of solvent and concentration dependence) and additives (scavengers, etc.). Based on such techniques to control reactions, we have recently obtained a vague picture of what is going on in the inorganic liquid systems.

The knowledge of the liquid phase reactions must also be important for interpreting the effects of dissolution on solids, and thus appear to be useful for more complete understanding of the solid phase hot atom chemistry. Moreover, hot atom chemistry of frozen solutions may provide us with information as to the state of aggregation of solutes in the frozen solvent matrices.

Solutions of Organometallic Compounds. Hot atom chemistry of organometallics has been studied mainly on solutions of metal cyclopentadienyls and carbonyls in organic solvents. Radicals play an important role on the reactions in these systems. Recoil species in these systems are generally separated by means of solvent extraction, sublimation or various chromatographic techniques. However, the spectrum of the products appears to be fairly complex and their assignments are generally incomplete. Accordingly, the reaction mechanisms have often been discussed in terms of 'retention' as the parent chemical form.

The processes eventually leading to the conservation of the original chemical form (i.e., retention) after the nuclear transformations in liquid phase may be classified as follows:

(i) "Apparent" primary retention.

(ia) Failure of bond rupture ('true' primary retention due to the momentum cancellation, or deexcitation without breaking the initial bonds).

(ib) Fast recombination of recoil fragments within the solvent cage (not scavengeable).

(ii) Secondary retention.

(iia) Substitution process (hot or epithermal substitution reactions with neighboring solute molecules).

(iib) Diffusive recombination process (slow recombination through thermal diffusive reactions of recoil fragments).

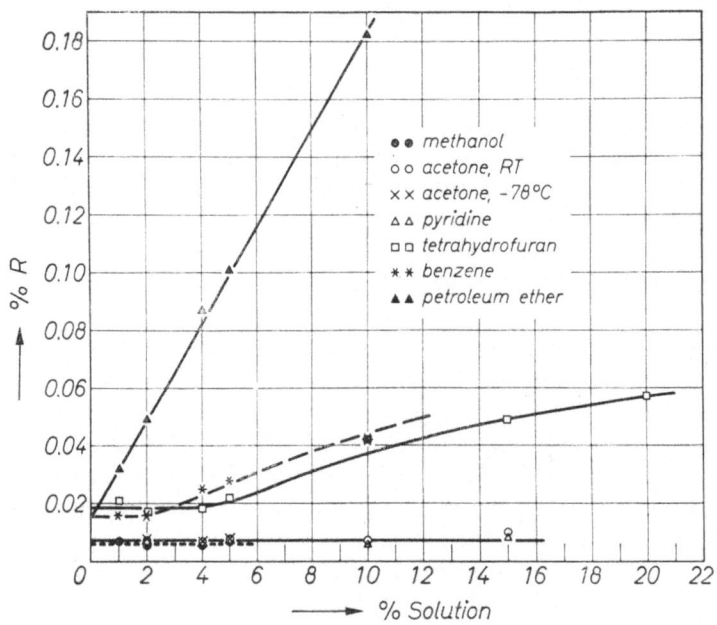

Fig. 3.2.5. The ^{56}Mn retentions observed in neutron-irradiated solutions of $C_5H_5Mn(CO)_3$ in various solvents (taken from Ref. 3.2.9)

The contributions of the processes (iia) and (iib) (i.e., secondary retention) can be estimated experimentally on the basis of the concentration dependence of the retention and the scavenger effect. Although the primary retention in definition should indicate the conservation of the original bonds of the recoil atom in the parent compound, it is difficult to make distinction between the processes (ia) and (ib) on experimental basis. Accordingly, the 'apparent' primary retention which can be obtained experimentally by suppressing the secondary retention includes overall contributions from these two processes, and in fact gives an upper limit to the 'true' primary retention. Accurate estimation of such an upper limit, which is generally a very small value, may provide useful information regarding the probability of bond rupture after nuclear transformations.

With a view to distinguishing between the various mechanisms leading to apparent retention, Zahn has investigated the concentration and temperature dependence of the ^{56}Mn retention after the neutron irradiation of dilute solutions of cyclopentadienyl manganese tricarbonyl $C_5H_5Mn(CO)_3$ and dimanganese decacarbonyl $Mn_2(CO)_{10}$ in various organic solvents [3.2.9]. In Fig. 3.2.5, the relatively small part of the ^{56}Mn retention, which is independent of concentration, temperature, or concomitant γ-dose in the reactor, is considered as of primary nature. Differences in the 'primary' retention values experimentally obtained in various solvents may likely be ascribed to a differing probability for a transfer of excitation energy from the excited $C_5H_5{}^{56}Mn(CO)_3$ molecule to the solvent molecules before bond rupture can take place. In strongly polar solvents such as methanol and acetone, the ^{56}Mn retention depends neither on concentration nor on temperature; this has been attributed to the suppression of thermal diffusive reactions by solvation of the recoil fragments. An alternative mechanism assumes that the radicals more easily formed from acetone or alcohol molecules in the vicinity of the recoiling atom may scavenge the recoil fragments and thus diminish the apparent retention [3.2.10]. Although the retention increases with the increase in concentration in non-polar solvents, it decreases quickly in the presence of a small amount of pyridine, probably due to the scavenging effect on thermal diffusive reactions. These observations may suggest that the concentration-dependent part of the ^{56}Mn retention in $C_5H_5Mn(CO)_3$ is mainly subjected to the reactions occurring at lower energies. By contrast the ^{56}Mn retention in dilute solutions of $Mn_2(CO)_{10}$ is independent of the nature of solvent molecules, and appears to be determined by the rupture of the weak Mn—Mn bond.

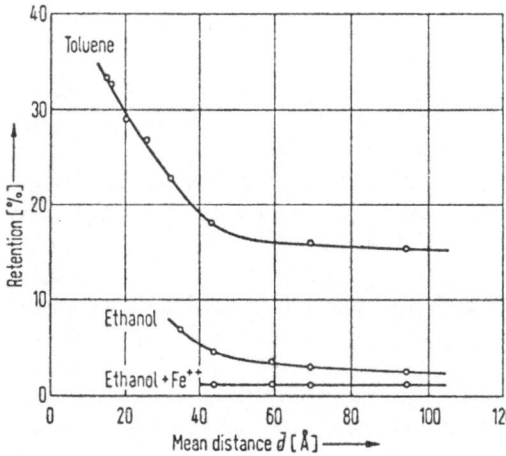

Fig. 3.2.6. Retention of ferrocene in toluene and ethanol solutions as a function of the mean distance between ferrocene molecules (taken from Ref. 3.2.10)

Lieser and Marcopoulos have studied the ^{59}Fe retention in irradiated organic solutions of ferrocene $Fe(C_5H_5)_2$ [3.2.10]. In Fig. 3.2.6, the ^{59}Fe retention values of ferrocene in toluene and ethanol solutions are plotted as a function of the mean distance of ferrocene molecules in the solutions calculated from the concentration. It is obvious that the ^{59}Fe retention of ferrocene depends on the mean distance (or concentration), kind of solvent, and scavengers. The retention curve for toluene solutions (Fig. 3.2.6) decreases sharply up to a mean distance of about 45 Å (or down to a concentration of about 0.02 M), and then tends to level off. As an explanation for such behavior, the region for the sharp decrease in retention has been ascribed to the substitution process since the probability for the recoiling ^{59}Fe atom to react with another ferrocene molecule should be higher at higher concentrations (or at smaller mean distances); at lower concentrations, the recombination of the recoil species with its former ligands, which is supposedly independent of dilution, becomes predominant. They have further suggested that the recoil range of ^{59}Fe in toluene solutions may be about 50 Å, on the assumption that the iron ions can react with ferrocene molecules only if these ions are in an excited state. The ^{59}Fe retention values at pseudo-plateau of the curve are appreciably lower in the ethanol solutions than in the toluene solutions; such a difference has been accounted for by assuming the interfering reactions of radicals radiolytically derived from ethanol with recombination of the recoil fragments. Similar scavenging effect was observed in the presence of glycerol. Complexing agents such as tartaric acid or citric acid also affect the recombination process. In the presence of 0.02 M Fe^{2+} ions as a scavenger, the retention is further lowered possibly because they compete with the ligand-deficient recoil fragments to combine with cyclopentadienyl radicals. However, the reaction mechanisms in the Fe^{2+}-containing ethanolic solutions may be somewhat complicated since there are more Fe^{2+} ions than ferrocene molecules in the system and recoil ^{59}Fe atoms arising from these scavenger Fe^{2+} ions also react with neighboring ferrocene molecules. The use of a non-isotopic scavenger, or ^{58}Fe-depleted iron salt, which produces practically no ^{59}Fe on neutron irradiation may be recommended for use to avoid such complexity.

In one of the papers on iron cyclopentadienyls and carbonyls by Wiles's group [3.2.11], ^{59}Fe(C$_5$H$_5$)$_2$ yields in Fe(C$_5$H$_5$)$_2$—Fe(CO)$_5$—C$_6$H$_6$ and ^{56}Fe(C$_5$H$_5$)$_2$—Fe(CO)$_5$—C$_6$H$_6$ were compared. On neutron irradiations at 30 °C, the ^{59}Fe(C$_5$H$_5$)$_2$ yield was 6.0 ± 0.3% in Fe(C$_5$H$_5$)$_2$—Fe(CO)$_5$—C$_6$H$_6$, and 4.5 ± 0.6% in ^{56}Fe(C$_5$H$_5$)$_2$—Fe(CO)$_5$—C$_6$H$_6$(^{56}Fe(C$_5$H$_5$)$_2$ was used as ^{58}Fe-depleted ferrocene); this reveals that recoil ^{59}Fe atoms resulting from neighboring Fe(CO)$_5$ mainly react with ferrocene molecules to yield the substituted product, ^{59}Fe(C$_5$H$_5$)$_2$. When frozen solutions were irradiated at −78 °C, however, the ^{59}Fe(C$_5$H$_5$)$_2$ yield was reduced by a factor of about 20 in the mixture containing ^{58}Fe-depleted ferrocene, implying that ferrocene was segregated as microcrystals in the solvent matrix.

Tominaga and coworkers have recently studied the effects of concentration and additives on the ^{60}Co retention in irradiated solutions of cobaltocene or a cobalticinium salts [3.2.12]. Since cobaltocene was easily oxidized to cobalticinium ion, the ^{60}Co yield as cobaltocene-cobalticinium ion (original structure being retained) was measured instead of the retention in cobaltocene benzene solutions; the ^{60}Co yield as cobaltocene-cobalticinium ion obviously depends on concentration of cobaltocene, and furthermore, is increased by addition of nickelocene as 'catcher' compound (Fig. 3.2.7). However, the ^{60}Co yield as cobaltocene-cobalticinium ion is not affected significantly by addition of ferrocene. These observations suggest that the ^{60}Co recoil atoms may react to form cobaltocene or cobalticinium ion in a lower energy range where the difference between chemical stabilities of these metallocenes can be distinguished. The ^{60}Co retention in the irradiated ethanolic or aqueous solutions of a cobalticinium salt was also concentration-dependent and partly scavengeable by iron(III) chloride.

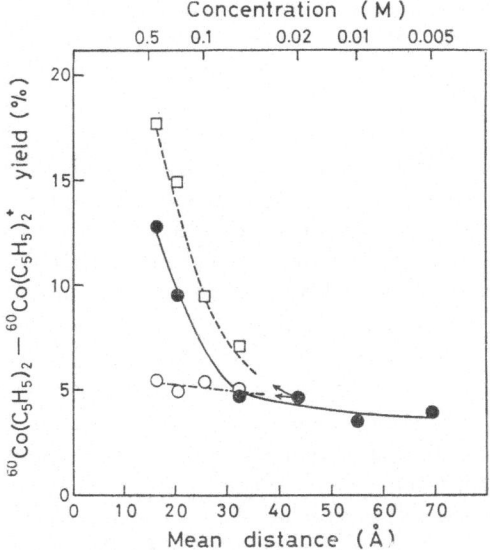

Fig. 3.2.7.
The ^{60}Co(C$_5$H$_5$)$_2$—^{60}Co(C$_5$H$_5$)$_2^+$ yield in irradiated solutions of cobaltocene in benzene (with/without additives) as a function of the mean distance between metallocene molecules (taken from Ref. 3.2.12).
— • —, cobaltocene solutions without additives; -- □ --, 0.02 M cobaltocene solutions containing 0.03, 0.08, 0.18, or 0.38 M nickelocene;
··· ○ ···, 0.02 M cobaltocene solutions containing 0.03, 0.08, 0.18, or 0.38 M ferrocene

The thermal diffusive reactions, which play an important role in recombination of the recoil species in the irradiated solutions of organometallics, appear to depend on various factors. Although the recoil products are only incompletely identified in most cases, it is likely that radicals arising from solvent molecules (e.g., phenyl radicals in benzene solutions [3.2.13]) are involved in the liquid phase recoil reactions; polymerization products (e.g., diferrocenyl or dicobaltocenyl [3.2.14]) may be less commonly formed in dilute solutions than in solid phase. Zahn has found that the contribution of the thermal diffusive reactions to the apparent retention decreases with increasing viscosity of the solvent, by comparing the ^{56}Mn retention in solutions of C$_5$H$_5$Mn(CO)$_3$ in various non-polar solvents [3.2.9]. Hillman and coworkers have reported that the retentions in solutions of metallocenes are lower than those of the phthalocyanines of the corresponding metals (Ti, Zr and Hf) [3.2.15, 16]. To account for such differences, the suggestion has been made that the recombination reactions of two bodies (a metal and a single organic moiety) for phthalocyanines would be more probable than the recombination of three bodies (the metal and two cyclopentadienyl groups) for metallocenes [3.2.17]. The suggestion has been confirmed by comparing the retentions of metallocenes (titanocene-, zirconocene-, and hafnocene dichloride) with those of their bridged derivatives (1,1′-trimethylenetitanocene dichloride and corresponding compounds of zirconium and hafnium, in which two cyclopentadienyl rings are tied together with a trimethylene bridge): in solutions, the bridged derivatives involving recombination of two bodies revealed higher retentions than the non-bridged parent compounds [3.2.18].

An earlier paper on organometallics has reported the reactions of fission product ruthenium with organic solvents [3.2.19]. When [103]Ru atoms were introduced into benzene through fission recoil and beta-decay, small yields of a [103]Ru-labeled organo-ruthenium compound were observed. However, when fission product [103]Ru or [106]RuCl$_3$ was mixed with cyclopentadiene, virtually all of the ruthenium was transformed into ruthenocene by a thermal reaction. Such a reaction may be of rather practical importance for synthesis of labeled ruthenocene.

Solutions of Transition Metal Complexes. Earlier studies on the hot atom chemistry of metal complexes in liquid phase were mostly concerned with the aqueous systems. The recoil ^{60}Co and [80]Br retentions have been reported for neutron-irradiated aqueous solutions of cobaltammine complexes such as Co(NH$_3$)$_6^{3+}$, Co(en)$_3^{3+}$ and Co(NH$_3$)$_5$Br^{2+} [3.2.20—22]. Various ^{51}Cr recoil species were identified in neutron-irradiated aqueous solutions of a chromium thiocyanate complex [3.2.23]. However, the aqueous solutions may not be appropriate systems for studying the detailed mechanisms of recoil reactions because of the strong interactions of the solvent (water) with the recoil species. Since water molecules easily coordinate most metal ions, ligand-deficient recoil species will be instantly stabilized through aquation in aqueous solutions; hence, the recombination toward parent via thermal diffusive reactions may seldom take place in aqueous systems unless the free ligand group has a stronger complexing power. Whenever aqueous solutions are exposed to reactor radiations, water molecules undergo radiation-chemical decomposition, and the radiolysis products may further react with the recoil species. The spectrum of the recoil products will be altered eventually by the secondary effect due to the medium. In view of such complexity arising from water as the solvent, we should better use non-polar organic solvents with high radiation stability (e.g., benzene), which only weakly interact with the recoil species. Tominaga and coworkers have investigated the hot atom reactions in irradiated solutions of some cobalt and chromium chelate complexes in such organic solvents by using various additives (scavengers, etc.) to distinguish between possible reaction mechanisms [3.2.24—30].

In neutron-irradiated solutions of tris(acetylacetonato)·cobalt(III), Co(acac)$_3$, in benzene, several percent of the total ^{60}Co activity produced remains as 'retention' in the parent chemical form ([60]Co(acac)$_3$). The retention is not practically concentration-dependent (or substitution mechanism was scarcely contributing), but increases slowly with time, revealing that the recombination proceeds through thermal diffusive reactions. The addition of an adequate metal salt (e.g., FeCl$_3$ or AlCl$_3$) prior to irradiation sharply lowers the retention down to nearly 0% (Fig. 3.2.8); the metal salt as scavenger competes with ligand-deficient ^{60}Co recoil species (e.g., bis(acetylacetonato)cobalt(II, or III), etc.) in thermal diffusive reactions to combine with free acetyl acetone, thus minimizing the [60]C retention effectively [3.2.24, 25]. As seen in Fig. 3.2.9, the apparent scavenging power of the metal salts correlates well with the thermodynamic stability constants of their acetylacetonates [3.2.25, 28]. It is obvious that the salts of

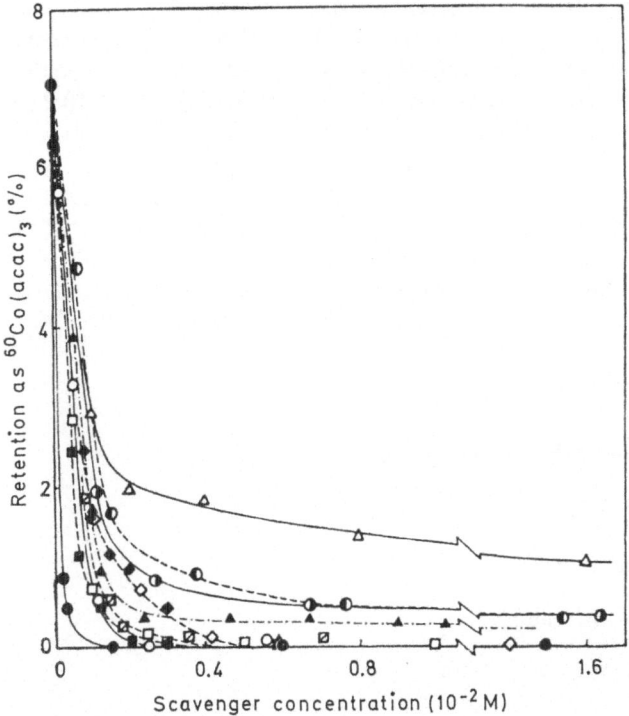

Fig. 3.2.8. Scavenger effect of various metallic salts on ^{60}Co retention in the neutron-irradiated 0.2 M solutions of Co(acac)$_3$ in benzene containing 10% of ethanol (taken from Ref. 3.2.25).

— ● — FeCl$_3 \cdot 6\,H_2O$ — ◐ — CoCl$_2 \cdot 6\,H_2O$
— ○ — FeCl$_3$ — ◖ — CoCl$_2$
— ▨ — AlCl$_3$ — △ — MgCl$_2 \cdot 6\,H_2O$
— ■ — CuCl$_2 \cdot 2\,H_2O$ — ◆ — HCl
— □ — Cu(CH$_3$CO$_2$)$_2 \cdot$ H$_2$O   — ◇ — HNO$_3$
— ▲ — NiCl$_2 \cdot 6\,H_2O$

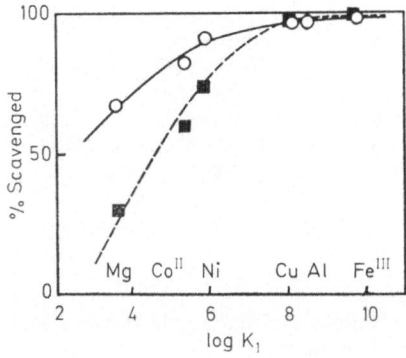

Fig. 3.2.9. Correlation of scavenging powers of various metal salts $(2 \times 10^{-3}\,M)$ with the thermodynamic stability constants of their acetylacetonates (Ref. 3.2.18). — ○ —: Separated within a few hours after irradiation. --- ■ ---: Separated after 7 days

metals which form more stable acetylacetonates (i.e., Fe(III), Al, Cu, etc.) work as effective scavengers. By contrast, the metal salts which form labile acetyl-acetonates (Mg, etc.) can only ineffectively scavenge the thermal reactions and the apparent ^{60}Co retention increases gradually with time (Fig. 3.2.10) [3.2.28]. Furthermore, additives such as Mg(acac)$_2$ (i.e., labile acetylacetonates) accelerate the increase in the apparent ^{60}Co retention remarkably, whereas stable acetyl-acetonates such as Fe(acac)$_3$ show practically no effect as additives (Fig. 3.2.10). Based on such results of the effects of additives, the overall reactions of the recoil species in solutions as well as the scavenging effect can be explained by postulating the combination of the following mechanisms [3.2.28, 30]:

(i) Recombination

$$[^{60}Co(acac)_{3-x}] + x\ acac^- \to {}^{60}Co(acac)_3$$

(ii) Scavenging reaction (MX$_n$: scavenger salt)

$$MX_n + n\ acac^- \to M(acac)_n$$

(iii) Ligand transfer reactions

$$[^{60}Co(acac)_{3-x}] + M(acac)_n$$

$$\to {}^{60}Co(acac)_3 + M(acac)_m \qquad (n > m).$$

The formula [^{60}Co(acac)$_{3-x}$] does not necessarily represent the actual chemical form of the ligand-deficient recoil species which might possibly be solvated (the electric charge on this species depends on the oxidation state of cobalt). Although the reactions at each step are fairly complicated, the above mechanisms have been confirmed at least for the final step leading to the parent chemical form (retention), ^{60}Co(acac)$_2 \to {}^{60}$Co(acac)$_3$, by simulating experiments, with photochemically produced species [3.2.31], or under various ambient atmospheres [3.2.32]. The ligand-deficient product formed by photolysis of Co(acac)$_3$ in benzene was found to undergo recombination reactions reforming Co(acac)$_3$. Such re-

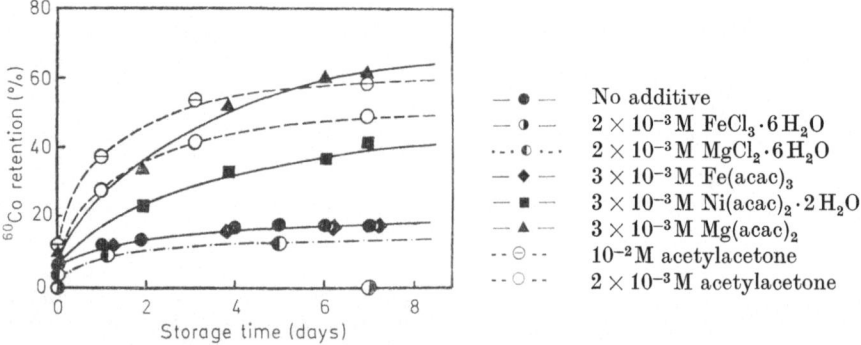

Fig. 3.2.10. ^{60}Co retention in the irradiated 0.2 M solutions of Co(acac)$_3$ in benzene containing 10% of ethanol and various additives (from Ref. 3.2.28).

combination reactions can be suppressed by addition of metal salts which form stable acetylacetonates and are transparent over the wavelength region of interest, indicating a striking resemblance to the scavenging effects observed with the recoil reactions in the similar systems [3.2.33].

Very similar results have been reported for the recoil reactions in organic solutions of tris(acetylacetonato)chromium(III) and tris(nitrosonaphtholato)cobalt(III) [3.2.26, 27, 29]. As has been demonstrated for the organometallics, the primary retention can be experimentally obtained as the minimum value in well-scavenged solutions. The 'apparent' primary retention thus estimated in benzene solutions of Co(acac)$_3$, Cr(acac), and tris (nitrosonaphtholato)cobalt(III) were 0.00 ± 0.02, 0.10 ± 0.04, and $1.1-1.6\%$, respectively. The different primary retention values may reflect the differences in the recoil energy spectrum, dexcitation efficiency and fast recombination mechanism within the cage, for different combination of metal-ligand-solvent.

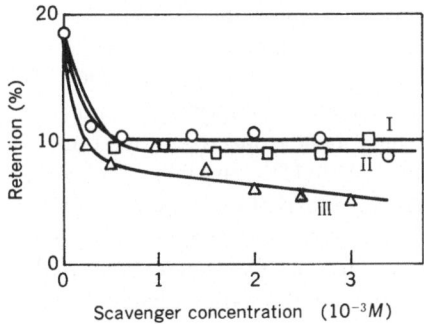

Fig. 3.2.11. ^{65}Ni retention in the neutron-irradiated 0.036 M solutions of bis(acetylacetone)ethylenediiminenickel(II) in Methylcellosolve as a function of the scavenger concentration (from Ref. 3.2.34). Scavengers: (I) FeCl$_2 \cdot 4\,$H$_2$O, (II) Co(CH$_3$COO)$_2$ $\cdot 4\,$H$_2$O, and (III) Cu(CH$_3$COO)$_2 \cdot$ H$_2$O

While the scavenging effects of metal salts in these systems are ascribed to the removal of free ligand groups through complex formation as mentioned above, there are other additives which can also scavenge thermal reactions by different mechanisms; acetate ions, for example, compete with free acetylacetonate groups to combine with the ligand-deficient recoil species, yielding products such as [^{60}Co(acac)$_2$(OOCCH$_3$)$_2$]$^-$ instead of ^{60}Co(acac)$_3$.

Ndiokwere and Elias have studied the ^{65}Ni retention in irradiated solutions of nickel complexes in various organic solvents [3.2.34]. The ^{65}Ni retention in neutron-irradiated solutions of bis(acetylacetone)ethylenediiminenickel(II) in Methylcellosolve was not dependent on concentration but was appreciably lowered by addition of ammonia, water, or metal salts (Fig. 3.2.11). This indicates that the retention consists of a primary retention of $5-6\%$, and a secondary retention mainly based on thermal recombination reactions. The scavenging mechanisms are explained by postulating that ammonia or water prevents recombination through coordination to the ligand-deficient ^{65}Ni recoil species, and that the metal salts suppress recombination via reactions with free ligand groups. As is seen in Fig. 3.2.11, the ^{65}Ni retention decreases quickly and approaches to a plateau in the presence of a cobalt(II) or iron(II) salt, while it continues to decrease gradually even after the sharp initial drop in the presence of a copper(II) salt. The anomaly with the Cu(II) has been ascribed to the replacement of ^{65}Ni by Cu in the parent molecules since the stability of the chelate complexes may increase in the order: Fe < Co < Ni < Cu. Hence, care must be taken, in choosing metal salts as scavengers, to avoid the use of metals which may form more stable complexes with the free ligand than the parent complex.

Only a few studies have been reported on the reactions of the recoil atoms arising from decay processes in liquid phase. The ^{80}Br retention following isomeric transition was studied in aqueous solutions of ^{80m}Br-labeled complexes $Co(NH_3)_5Br^{2+}$ and $PtBr_6^{2-}$ [3.2.35]. While the $Co(NH_3)_5{}^{80m}Br^{2+}$ ions were completely disrupted after isomeric transition in solutions possibly because of Coulombic explosion, nearly 50% of the $PtBr_5{}^{80m}Br^{2-}$ ions remained unaltered even after the decay process. Both substitution and recombination reactions are unlikely to take place in the $PtBr_5{}^{80m}Br^{2-}$ solutions, in view of the lack of concentration dependence and the susceptibility of the complex to aquation. Accordingly, the high ^{80m}Br-retention for the hexabromoplatinate(IV) ions has been explained by assuming that negatively charged complex ions can more easily tolerate the increase in positive charge of ^{80}Br caused by the Auger process after isomeric transition. A similar study on the retention of ^{58m}Co-labeled Co(III) complexes after isomeric transition has been reproted recently [3.2.36]. The probability that a complex can survive a violent decomposition (Coulombic explosion) due to inner shell ionization of the cobalt atom after the transition $^{58m}Co \rightarrow {}^{58}Co$ was investigated for various cobalt(III) ammine- and cyano complexes such as $[Co(NH_3)_6](NO_3)_3$, $[Co(en)_3]X_3$, and $K_3[Co(CN)_6]$. The survival probability, or the ^{58}Co retention, apparently depends upon the chemical constitution of the complexes both in solid state and solutions. The survival probability of $Co(acac)_3$ in solutions is also dependent on the kind of organic solvents, ranging from 0.3% in $(CH_3)_2S$ to 8.7% in $CHCl_3$.

The chemical effects following β-decay have been studied in aqueous solutions of ^{144}Ce- or ^{143}Ce-labeled cerium complexes [3.2.37, 38]. Since isotopic exchange and substitution reactions are often involved, solution chemistry of such compounds should have been known reasonably well for complete understanding of the overall mechanisms of the reactions of recoil species in such systems.

Frozen Solutions. Frozen solutions are interesting systems for hot atom chemistry because of their intermediate character between liquid and solid phases.

On freezing solutions, the solutes often form clusters or microcrystals heterogeneously in the solvent matrices. Since the size of reaction cages for recoiling atoms in condensed phases is generally small as compared with the dimension of the aggregates, their formation may appreciably alter the yields of recoil products. Hence, the concentration dependence of the yields in frozen solutions can give useful information regarding the state of aggregation of the solutes. In fact, clustering of halogens has been observed in frozen solutions of halogens in organic solvents by employing such technique [3.2.39—43].

The retentions reported for the frozen solutions of $Ru(C_5H_5)_2$, $C_5H_5Mn(CO)_3$ and $Fe(C_5H_5)_2$ are fairly close to those for their solids, demonstrating that the solutes may be segregated as microcrystals in the frozen state over the concentration range studied [3.2.9, 11, 44, 45].

Tominaga has investigated the recoil ^{60}Co reactions of Co(acac)$_3$ both in heterogeneous solution-sloid mixtures and in frozen solutions [3.2.46]. The ^{60}Co retentions in the mixtures of a saturated acetic acid solution with various amounts of solid Co(acac)$_3$ agree well with the values calculated by simply combining the ^{60}Co retentions in the solution and solid phase (Fig. 3.2.12a). The concentration dependence of the ^{60}Co retention in the frozen acetic acid solution of this complex indicates a very similar behavior (Fig. 3.2.12b), and suggests that a homogeneous frozen solution can only be obtained below a critical concentration of about 10^{-3} M.

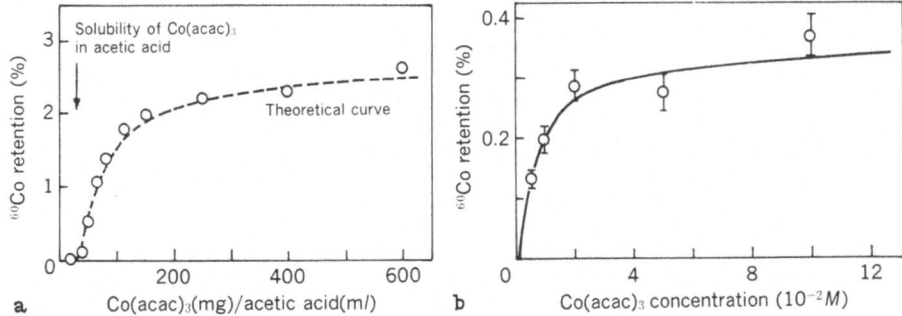

Fig. 3.2.12. a ^{60}Co retention vs. the ratio Co(acac)$_3$/acetic acid in the irradiated acetic acid solutions and solution-solid mixtures. **b** ^{60}Co retention vs. the Co(acac)$_3$ concentration in the irradiated frozen Co(acac)$_3$ — acetic acid systems (taken from Ref. 3.2.46)

References

[3.2.1]   Frank, J., Rabinowitch, E.: Trans. Faraday Soc. *30*, 120 (1934)
[3.2.2]   Richardson, A. E., Wolfgang, R.: J. Am. Chem. Soc. *92*, 3480 (1970)
[3.2.3]   Arsenalt, L. J., Blotcky, A. J., Medina, V. A., Rack, E. P.: J. Phys. Chem. *83*, 893 (1979)
[3.2.4]   Berg, M. E., Loventhal, A., Adelman, D. J., Grauer, W. M., Rack, E. P.: ibid. *81*, 837 (1977)
[3.2.5]   Machulla, H. J., Stocklin, G.: J. Phys. Chem. *78*, 658 (1974)
[3.2.6]   Vararos, L., Machulla, H. J. Stöcklin G.: J. Phys. Chem. *76*, 501 (1972)
[3.2.7]   Stöcklin, G.: *Chemie heißer Atome*. Weinheim: Verlag Chemie 1969
[3.2.8]   Wolf, A. P., Schueler, P., Pettijohn, R. P., To, K. C., Rack, E. P.: J. Phys. Chem. *83*, 1237 (1979)
[3.2.9]   Zahn, U.: Radiochim. Acta *7*, 170 (1967)
[3.2.10] Lieser, K. H., Marcopoulos, Chr.: Radiochim. Acta *22*, 111 (1975)
[3.2.11] Kanellakopoulos, W., Wiles, D. R.: Radiochim. Acta *16*, 179 (1971)
[3.2.12] Sato, H., Yoshida, M., Tominaga, T.: Radiochem. Radioanal. Lett. *44*, 159 (1980)
[3.2.13] Yang, I. C., Wiles, D. R.: Canadian J. Chem. *45*, 1357 (1967)
[3.2.14] Wheeler, O. H., McClin, M. L.: Radiochim. Acta *7*, 181 (1967)
[3.2.15] Hillman, M., Kim, C. K., Shikata, E., Weiss, A. J.: Radiochim. Acta *9*, 212 (1968)
[3.2.16] Hillman, M., Weiss. A. J., Ebihara, H., Williams, K.: J. Inorg. Nucl. Chem. *37*, 403 (1975)
[3.2.17] Hillman, M., Weiss, A. J., Hahne, R. M. A.: Radiochim. Acta *12*, 220 (1969)
[3.2.18] Hillman, M., Weiss, A. J.: J. Inorg. Nucl. Chem. *39*, 1921 (1977)
[3.2.19] Zahn, U., Harbottle, G.: J. Inorg. Nucl. Chem. *28*, 925 (1966)

[3.2.20] Kayas, G., Sue, P.: J. Chim. Phys. *45*, 188 (1948)
[3.2.21] Zuber, A.: USAEC Doc. NYO-6142 BNL (1954)
[3.2.22] Saito, N., Sano, H., Tominaga, T.: Bull. Chem. Soc. Jpn. *33*, 20 (1960)
[3.2.23] Kaufman, S.: J. Am. Chem. Soc. *82*, 2963 (1960)
[3.2.24] Tominaga, T., Fujiwara, K.: Bull. Chem. Soc. Jpn. *43*, 2279 (1970)
[3.2.25] Tominaga, T., Sakai, T., Fujiwara, K.: Bull. Chem. Soc. Jpn. *44*, 3036 (1971)
[3.2.26] Tominaga, T., Nishi, Y.: Radiochem. Radioanal. Lett. *8*, 151 (1971)
[3.2.27] Tominaga, T., Sakai, T.: Bull. Chem. Soc. Jpn. *45*, 1237 (1972)
[3.2.28] Tominaga, T., Nishi, Y.: Bull. Chem. Soc. Jpn. *45*, 3213 (1972)
[3.2.29] Tominaga, T., Nishi, Y.: Radiochem. Radioanal. Lett. *11*, 289 (1972)
[3.2.30] Nishi, Y., Tominaga, T.: Radioisotopes *23*, 700 (1974)
[3.2.31] Nishi, Y., Tominaga, T.: Radiochem. Radioanal. Lett. *24*, 249 (1976)
[3.2.32] Tominaga, T., Nishi, Y., Motohashi, E.: Radiochem. Radioanal. Lett. *18*, 15 (1974)
[3.2.33] Tominaga, T., Nishi, Y.: Radiochem. Radioanal. Lett. *31* 129 (1977)
[3.2.34] Ndiokwere Ch. L., Elias, H.: Radiochim. Acta *19*, 181 (1973)
[3.2.35] Adamson, A. W., Grunland, J. M.: J. Am. Chem. Soc. *73*, 5508 (1951)
[3.2.36] Lazzarini, E., Lazzarini, A. L. F.: J. Inorg. Nucl. Chem. *39*, 207 (1977)
[3.2.37] Glentworth, P., Wiseall, B.: Chem. Effects Nucl. Transform., Proc. Symp. Vienna, 1965, Vol. II, p. 483
[3.2.38] Shiokawa, T., Omori, T.: Bull. Chem. Soc. Jpn. *38*, 1340, 1892 (1965); *42*, 696 (1969); *43*, 2076 (1970)
[3.2.39] Iyer, R. M., Willard, J. E.: J. Am. Chem. Soc. *88*, 4561 (1966)
[3.2.40] Abedinzadeh, Z., Grillet, S., Stevovic, J., Tanaka, K., Milman, M.: Radiochim. Acta *9*, 38 (1968)
[3.2.41] Ayres, R. L., Kemnitz, E. J. Lambrecht R. M., Rack, E. P.: Radiochim. Acta *11*, 1 (1969)
[3.2.42] Lambrecht, R. M., Hahn, H. K. J., Rack, E. P.: J. Phys. Chem. *73*, 2779 (1969)
[3.2.43] Kemnitz, E. J., Hahn, H. K. J., Rack, E. P.: Radiochim. Acta *13*, 112 (1970)
[3.2.44] Harbottle, G., Zahn, U.: Chem. Effects Nucl. Transform. Proc. Symp. Vienna, 1965, Vol. II, p. 133
[3.2.45] Zahn, U.: Radiochim. Acta *8*, 177 (1967)
[3.2.46] Tominaga, T.: Radioisotopes *22*, 411 (1973)

3.3 Solid Phase Hot Atom Reactions

Since its application to production of highly enriched radioisotopes in the early days, hot atom chemistry in the solid phase has been studied extensively on a number of inorganic compounds such as oxyacids and metal complexes [3.3.1—4]. These solid phase studies have mostly employed the 'classical' or conventional approach to hot atom chemistry, i.e., dissolution of irradiated solids, chemical separations, and determination of yields and annealing effects [3.3.5]. However, none of the reaction models so far proposed to account for experimental data have proved to be satisfactory for complete understanding of solid phase recoil chemistry.

With such conventional approach, the real picture of the original solid phase phenomenon may inevitably be altered by annealing reactions and chemical reactions induced during and after dissolution of the solid. Although physical techniques such as Mössbauer spectroscopy have recently become applicable to in-situ analysis of recoil species [Sect. 2.3.2], they can only provide limited amount of information regarding their chemical identities. Hence, the conventional approach still remains complementary to the physical means, if it is carefully adapted in well controlled experiments. However, a number of 'classical' works so far reported in the solid state may now better be evaluated as applications in production of labeled compounds [see Sect. 4.1] rather than as mechanistic studies of recoil reactions.

While consequences of the nuclear recoil after relaxation of excess charge and kinetic energy (in a period of time 10^{-9} to 10^{-7} s, or longer term period) can be studied by either these physical or chemical techniques, very little is known of the slowing down processes for the recoiling atoms (in a time period of $\sim 10^{-12}$ s) in the solid state. The theoretical approach (e.g., computer calculations) to the latter problem would require detailed basic knowledge on the properties of solids, and collaboration with solid state physicists.

Appearance Energy

The atoms which have acquired large kinetic energies by nuclear recoil produce fragmentation on the molecules and migrate to interstitial positions in solids. In order to estimate the minimum energy required for such displacement in the solid state, Yoshihara and coworkers have studied the radiochemically separable yields in indium and lutetium complexes as a function of the recoil energy [3.3.6—8]. Indium-edta complex was irradiated with 2—8 MeV γ-rays to induce $^{115}In(\gamma, \gamma')^{115m}In$ reaction. The recoil ^{115m}In atoms released from the complex after bond rupture were separated as $^{115m}In^{3+}$ ions by cation exchange. In Fig. 3.3.1, the radiochemical yield of $^{115m}In^{3+}$ is plotted against the recoil energy attained mainly through (γ, γ') reaction [3.3.6]. For energies below 60 eV the separable yield remains at nearly 0%, indicating that the recoil ^{115m}In atoms lack energies to break bonds and/or to escape recombination with the former ligands. The yield increases abruptly at 60—70 eV and then levels off.

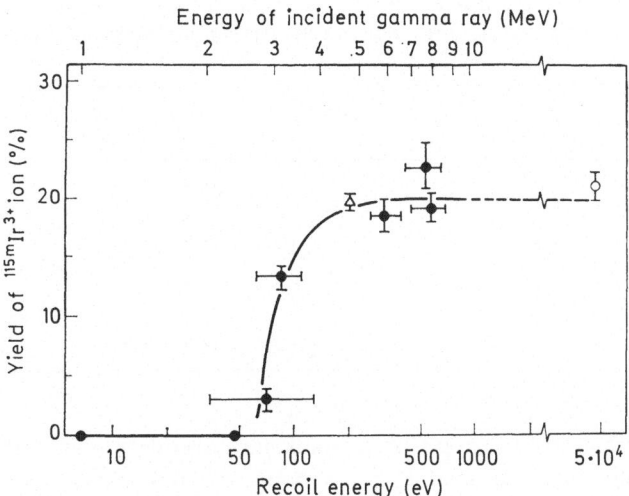

Fig. 3.3.1. Yield of $^{115m}In^{3+}$ ion following the $^{115}In(\gamma, \gamma')^{115m}In$ reaction in the In-edta complex as a function of the recoil energy (taken from Ref. 3.3.6). ●, $^{115}In(\gamma, \gamma')^{115m}In$ reaction; △, $^{115}In(n, \gamma)^{116m}In$ reaction; ○, $^{115}In(\gamma, n)^{114m}In$ reaction

The appearance energy, or the energy required for fragmentation without being recombined, depends upon the parent complex, e.g., 60 eV for In-edta and 40 eV for Lu-edta, probably due to the difference in recombination processes between the recoil atom and ligand [3.3.6, 7]. Similar dependence of the separable yield on the recoil energy has also been reported for zinc phthalocyanine [3.3.9].

Mechanisms of Solid Phase Hot Atom Reactions

The charge spectrometry of decayed atoms in the gas phase indicates that they have acquired large positive charges shortly after nuclear events [Sect. 2.3.1]. Although formation of highly charged atoms can also be expected as the effects of nuclear transformations in the solid state, this has been observed experimentally only for the decaying atoms embedded on the surface of solids [3.3.10]. The relaxation of excessive positive charge is so rapid in metals that we observe no 'anomalous' charge states after the nuclear transformations in such solids. However, slightly oxidized states can be detected by the Mössbauer emission technique in some insulators where the highly positive charge has been neutralized incompletely within a time period of 10^{-9} to 10^{-7} s. For example, ^{57}Fe atoms produced in the decay of ^{57}Co-labeled cobalt(II) halides assume the same oxidation state as in the parent compound ($+2$) for the chloride, bromide, and iodide, whereas they are found partly in the oxidized state ($+3$) for the fluoride. The difference has been explained by postulating that the anomalous ^{57}Fe state ($+3$) can be further neutralized by electrons from the coordinating halogens (Cl, Br, I) with large polarizabilities [3.3.11, 12]. In general, the Mössbauer studies suggest that in 10^{-9} to 10^{-7} s after the nuclear transformations the decayed atoms have been nearly stabilized in the oxidation states in accordance with their chemical environment.

Table 3.3.1. Yield of ^{57}Fe(II) after EC decay (Mössbauer), G(Co^{2+}) value in radiolysis and radiochemical yield of ^{60}Co(II) from (n, γ) reaction in cobalt(III) complexes containing oxalate ion(s) (taken from Ref. 3.3.15)

Compounds	Yield (%) of ^{57}Fe(II) in EC decay	G(Co^{2+})	Yield (%) of ^{60}Co(II) in (n, γ) reaction
[Co(NH$_3$)$_6$](NO$_3$)$_3$	40 ± 1	1.1 ± 0.2	66.7 ± 0.2
[Co(en)$_3$](NO$_3$)$_3$	53 ± 1	0.4 ± 0.1	69.4 ± 0.2
[Co(NH$_3$)$_6$]$_2$(C$_2$O$_4$)$_3$4 H$_2$O	72 ± 1	2.9 ± 0.2	91.0 ± 0.2
[Co(en)$_3$]$_2$(C$_2$O$_4$)$_3$9 H$_2$O	74 ± 1		
[Co(NH$_3$)$_6$][Cr(C$_2$O$_4$)$_3$]3 H$_2$O	59 ± 1		
[Co(NH$_3$)$_6$][Fe(C$_2$O$_4$)$_3$]3 H$_2$O	68 ± 1		
K$_3$[Co(C$_2$O$_4$)$_3$]3 H$_2$O	85	11.6 ± 0.2	98.6 ± 0.4

Oxidation states lower than in the parent compounds have also been observed by Mössbauer spectroscopy in some solid systems. Sano and coworkers have studied the chemical effects of EC-decay in a number of ^{57}Co-labeled Co(III) complexes and found formation of ^{57}Fe(II) in such systems [3.2.13—16]. They have suggested that these chemical effects can be accounted for in terms of local radiolytic processes initiated by the Auger effect following the EC-decay. Table 3.3.1 summarizes the $^{57}Fe(II)$ yield from EC-decay in ^{57}Co-labeled Co(III) complexes containing oxalate ion(s), together with the $G(Co^{2+})$ value in radiolysis of the same complex. It is obvious that the oxalate ion(s) directly coordinated to cobalt shows marked reducing effect on the $^{57}Fe(II)$ yield, and that the oxalate ion(s) in the second coordination sphere is also slightly reductive. The reducing effect of the oxalate ion may be ascribed to electrons released in its radiolytic decomposition induced by the Auger effect,

$$C_2O_4^{2-} \rightarrow 2CO_2 + 2e^-$$

as has been reported for pyrolysis, photolysis and γ-radiolysis of metal oxalates [3.3.17—19]. The $^{57}Fe(II)$ yield was found to be larger in $[^{57}Co(NH_3)_6][Fe(C_2O_4)_3]$ $\cdot 3H_2O$ than in $[^{57}Co(NH_3)_6][Cr(C_2O_4)_3]\cdot 3H_2O$: the difference was related to the fact that $K_3[Fe(C_2O_4)_3]\cdot 3H_2O$ suffers more decomposition than $K_3[Cr(C_2O_4)_3]\cdot 3H_2O$ in γ-radiolysis, in terms of the $G(CO_2)$ value [3.3.20]. These observations may indicate that the reactive zone produced by the EC and Auger effects includes the neighboring $[M(C_2O_4)_3]^{3-}$ ions completely, or is as large as 20 Å in diameter [3.3.21].

The size of such reactive zone in solids has also been demonstrated in another experiment with metal acetylacetonates [3.3.13]. A comparison was made of the Mössbauer spectra taken for the γ-radiolysis in ^{57}Fe-doped $Mn(acac)_3$, $Fe(acac)_3$ and $Co(acac)_3$, and for the ^{57}Co EC-decay in ^{57}Co-doped $Mn(acac)_3$, $Fe(acac)_3$ and $Co(acac)_3$, as illustrated in Fig. 3.3.2 [3.3.13]. If the Auger-effect-induced reactions involved only the ligand acetylacetonate ions directly bonded to ^{57}Co, the spectra 2a, c and e should be all identical or at least indicate nearly identical yields of $^{57}Fe(II)$. In fact, the $^{57}Fe(II)$ yield clearly depends on the host material, indicating that the radiolytic processes following the Auger effect are governed not only by the ligands in the first coordination sphere but also by the neighboring metal acetylacetonate molecules (host material). Furthermore, the striking resemblance of the emission Mössbauer spectra for the ^{57}Co EC-decay to the Mössbauer absorption spectra for the γ-radiolysis of $^{57}Fe(acac)_3$ doped in the similar host materials may reveal that the EC-Auger effects can be generally understood as the local radiolytic processes [3.3.13].

The Mössbauer technique may also provide information regarding the radiolytic effect produced in the surroundings as well as the chemical states of the Mössbauer atoms themselves. The paramagnetic relaxation, and the peak intensity and the line width, in the Mössbauer spectrum of the decaying atom are useful as the probe for detecting the damages produced in the neighboring matrix [3.3.21]. Paramagnetic solids in general show no magnetic splitting in the Mössbauer spectra: the $^{57}Fe(III)$ nucleus experiences no magnetic field on average bacause of the rapid fluctuations in the internal magnetic field due to electronic spins. If the paramagnetic $^{57}Fe(III)$ species is isolated by dilution in diamagnetic materials, the relaxation time becomes longer, causing the magnetic splitting in the spectra because the spin-spin interaction with the remote paramagnetic species is weakened. In fact, the $^{57}Fe(acac)_3$ isolated by dilution in diamagnetic $Co(acac)_3$ obviously demonstrates such paramagnetic hyperfine structure [3.3.13]. However, the spectrum of ^{57}Co-labeled $Co(acac)_3$ (Fig. 3.3.2a) (i.e., $^{57}Fe(III)$

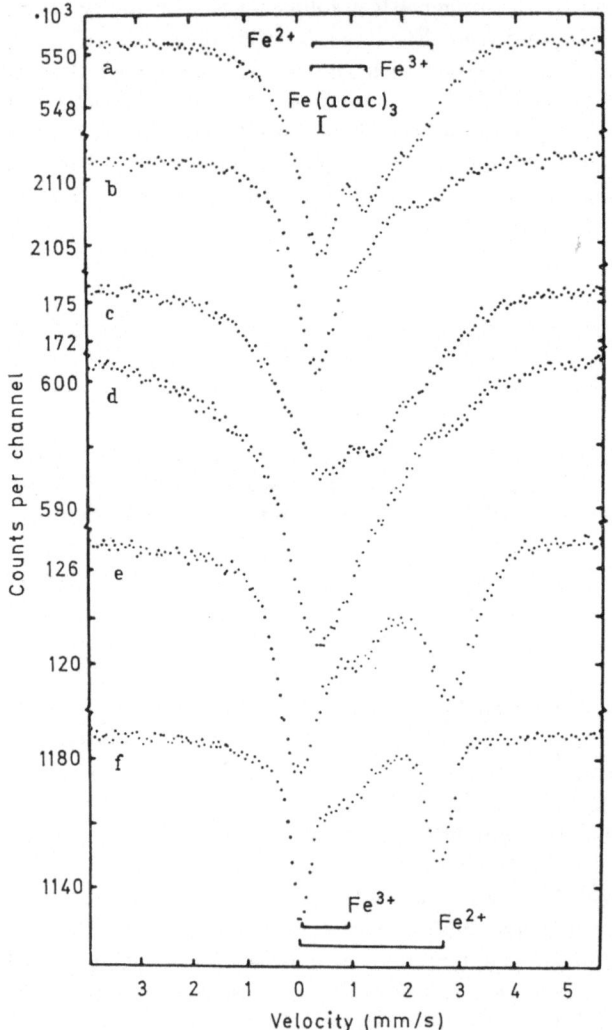

Fig. 3.3.2. Mössbauer spectra at 78 K of metal acetylacetonates (taken from Ref. 3.3.13).
a) ^{57}Co-labeled Co(acac)$_3$ source
b) γ-irradiated (3.0×10^9 R) (Co, ^{57}Fe)(acac)$_3$ absorber
c) (Mn, ^{57}Co)(acac)$_3$ source
d) γ-irradiated (3.0×10^9 R) (Mn, ^{57}Fe)(acac)$_3$ absorber
e) (Fe, ^{57}Co) source
f) γ-irradiated (3.0×10^9 R) Fe(acac)$_3$ absorber.
All the velocity scales were normalized with respect to iron

isolated in a diamagnetic matrix) never shows paramagnetic relaxation. The difference has been ascribed to a shorter relaxation time in the latter system due to the interaction of ^{57}Fe(III) with the electronic spins of radicals formed in its vicinity [3.3.13]. Thus, the disappearance of paramagnetic relaxation may provide an evidence for formation of radicals in the surroundings of the decaying atoms [3.3.21]. These radicals will end up by forming CO, CO_2, H_2, CH_4, CH_3COCH_3, or C_2H_6 molecules as observed gas chromatographically in γ-radiolysis of the metal acetylacetonates [3.3.22].

The lattice dynamic information for surroundings of the Mössbauer atoms may be reflected on the peak intensity or the line width in the spectra. The recoil-free fraction, derived by comparing temperature dependence of the peak intensity, can be taken as a measure for estimating how rigidly the Mössbauer atom is bound in the solid. If the neighboring lattice has been destroyed by the decaying Mössbauer atom, the recoil-free fraction becomes smaller (or peak intensity is quickly decreased with the increase in temperature) because of the increased mobility, and the line width is broadened because of the overlapping of slightly different chemical states produced in the disturbed environment. For example, the larger peak intensity of ^{237}Np formed in β-decay than that in α-decay may reveal the greater after-effect in the latter nuclear event [3.3.23, 24].

Effects of Dissolution of Solids. It has been believed that 'classical' results on the chemical states of recoil atoms in solids generally depend on the experimental conditions such as the time and temperature of irradiation and storage, the chemical properties of solvent, and the type of radiochemical separation techniques employed [3.3.25—27]. Now we may raise this question: are all 'classical' works on the solid state obtained by radiochemical means of analysis simply a pile of useless data for mechanistic studies?

In this respect, Sano and coworkers have recently reported an interesting work in which they compared the yields of ^{57}Fe(II) species in the EC-decay of ^{57}Co with those of ^{60}Co(II) species produced in the ^{59}Co(n, γ)^{60}Co reactions in twelve cobalt(III) ammine complexes (Fig. 3.3.3) [3.3.16]. The ^{57}Fe(II) yield obtained by Mössbauer technique correlates fairly well with the ^{60}Co(II) yield obtained by classical radiochemical processes [3.3.28, 29], except for $[Co(NH_3)_6]_2(CrO_4)_3$

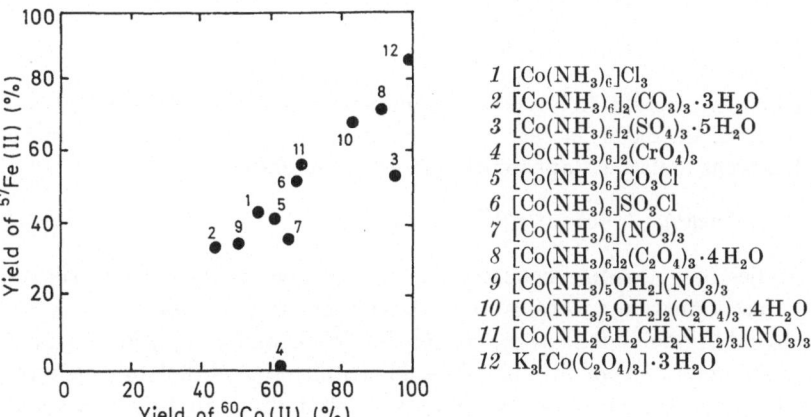

1 $[Co(NH_3)_6]Cl_3$
2 $[Co(NH_3)_6]_2(CO_3)_3 \cdot 3\,H_2O$
3 $[Co(NH_3)_6]_2(SO_4)_3 \cdot 5\,H_2O$
4 $[Co(NH_3)_6]_2(CrO_4)_3$
5 $[Co(NH_3)_6]CO_3Cl$
6 $[Co(NH_3)_6]SO_3Cl$
7 $[Co(NH_3)_6](NO_3)_3$
8 $[Co(NH_3)_6]_2(C_2O_4)_3 \cdot 4\,H_2O$
9 $[Co(NH_3)_5OH_2](NO_3)_3$
10 $[Co(NH_3)_5OH_2]_2(C_2O_4)_3 \cdot 4\,H_2O$
11 $[Co(NH_2CH_2CH_2NH_2)_3](NO_3)_3$
12 $K_3[Co(C_2O_4)_3] \cdot 3\,H_2O$

Fig. 3.3.3. Comparison of the radiochemical yield of ^{60}Co(II) species produced in the ^{59}Co(n, γ)^{60}Co hot-atom reactions with the yield of ^{57}Fe(II) species produced after EC-decay (from Ref. 3.3.16)

which shows an anomalously high ^{60}Co(II) yield for its low ^{57}Fe(II) yield. The correlation between the ^{60}Co(II) and the ^{57}Fe(II) yields demonstrates the similarity in the fates of the ^{60}Co atoms arising from the (n, γ) reaction and the ^{57}Fe atoms produced in the EC-decay, also implying that the general trend for chemical distribution of recoil ^{60}Co atoms in solids may not always be influenced significantly by conventional radiochemical separation processes involving dissolution of the solids. The anomalous result for $[Co(NH_3)_6]_2(CrO_4)_3$ has been attributed to the oxidative character of CrO_4^{2-} ions, which will either favor the Fe(III) state for the decayed ^{57}Fe atoms by picking up electrons through the process,

$$CrO_4^{2-} + 2e^- \rightarrow CrO_4^{4-}$$

or consume ligands NH$_3$ groups to yield ligand-deficient recoil species: both ligand-deficient ^{60}Co(II)- and ^{60}Co(III) species may decompose to a reduced form on dissolution in water and apparently increase the ^{60}Co(II) yield by the classical methods [3.3.16]. Hence, it is obvious that the results of the classical hot atom chemistry correlates reasonably well with those obtained with the Mössbauer emission technique unless any serious perturbation takes place during the chemical analysis of the dissolved solid. Such comparison between the classical and physical studies may be useful for understanding of the detailed mechanisms of hot atom reaction in the solid phase.

Generally speaking, the results of hot atom chemistry in solids by classical methods may reveal combined effects of solid phase reactions and subsequent liquid phase reactions on and after dissolution of the solids. In order to acquire a true picture of solid phase hot atom reactions we need to make corrections for the subsequent reactions in solutions, or to develop experimental techniques for minimizing such dissolution effects. In polar solvents such as water, ligand-deficient recoil species are either stabilized instantly by solvation,

$$^{60}Co^{III}(NH_3)_5^{3+} + H_2O \rightarrow {}^{60}Co^{III}(NH_3)_5H_2O^{3+},$$

or decomposed to yield other species,

$$^{60}Co^{II}(NH_3)_5^{2+} + H_2O \rightarrow {}^{60}Co_{aq}^{2+}$$

inhibiting further reaction. In less polar organic solvents such as benzene, however, the ligand-deficient recoil species may recombine with former ligands rather than with solvent molecules, apparently increasing retention,

$$^{60}Co^{III}(acac)_2^+ + acac^- \rightarrow {}^{60}Co^{III}(acac)_3.$$

Recent studies of hot atom chemistry in organic solvents [Sect. 3.2.2] suggest that such recombination processes after dissolution can be suppressed experimentally by addition of adequate scavengers for free ligand groups. For instance, the apparent retention of irradiated solid tri(nitrosonaphtholato)cobalt(III) dissolved in pure benzene is nearly 85%, whereas it drops to 15% if a cobalt or copper salt is present as a scavenger (Fig. 3.3.4) [3.3.30]. The latter value may be regarded as the true solid-phase retention, while the difference (70%) should be ascribed to the recombination processes in solution. In view of such striking effect some of the classical retention values of solid chelate complexes should be reexamined by the scavenger technique since they could have been influenced appreciably by the recombination processes in solution. In the presence of adequate scavengers to suppress thermal reactions, reasonably low retention values have been obtained reproducibly for chelate complexes (e.g., Co(acac)$_3$ and Cr(acac)$_3$) irrespective of experimental conditions such as the time after dissolution, the kind of solvent, the type of analytical technique, or the concentration of other impurities [3.3.31, 32]. Thus the *minimum* retention value obtained by scavenger technique may likely represent the *true* retention in the *solid* state.

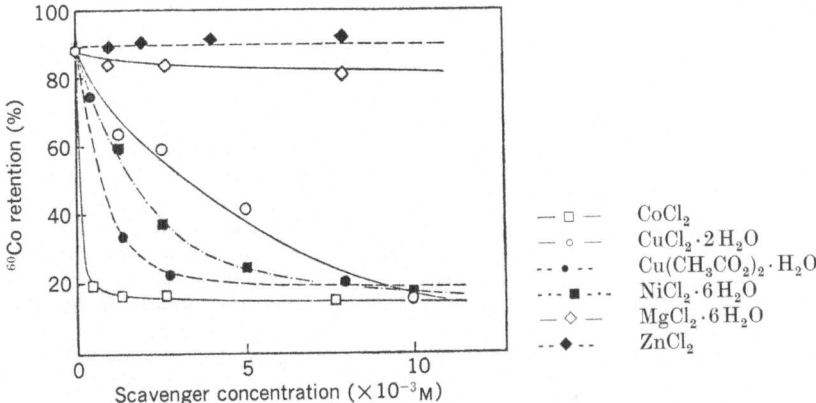

Fig. 3.3.4. Scavenger effect of various metallic salts on ^{60}Co retention in the 0.01 M solutions of irradiated solid tris(α-nitroso-β-naphtholato)-cobalt(III) (taken from Ref. 3.3.30)

The comparison between the behavior of the recoil species in solid phase and that on dissolution in organic solvents may be necessary for clarifying their reaction mechanisms. The ^{51}Cr retention has been found to increase with time after dissolution when irradiated solid Cr(acac)$_3$ was dissolved and heated in benzene or methanol: this phenomenon apparently resembles the increase in retention of the irradiated solid on heating, or thermal annealing behavior (Fig. 3.3.5) [3.3.33]. By contrast, the radiochemical yield of ligand-deficient ^{51}Cr species (the fraction elutable with methanol from an alumina column in Fig. 3.3.5) has been decreased with time on heating the solution. These observations demonstrate that the ligand-deficient species produced in solids recombine with free ligand groups in solutions [3.3.34]. However, the apparent similarity between the time-dependent behavior of retention in solution and the annealing behavior in the solid state may not necessarily indicate the essential similarity between the mechanisms of reactions of the recoil species in solution and in solid. For example, the atmospheric oxygen inhibits the thermal annealing in irradiated Co(acac)$_3$ [3.3.35], whereas it enhances the increase in the retention of the same compound after dissolution in an organic solvent [3.3.36]. A dimeric recoil species has been proposed as the product from the hot atom reactions in solid Co(acac)$_3$ [3.3.37], and formation of polynuclear complexes has been

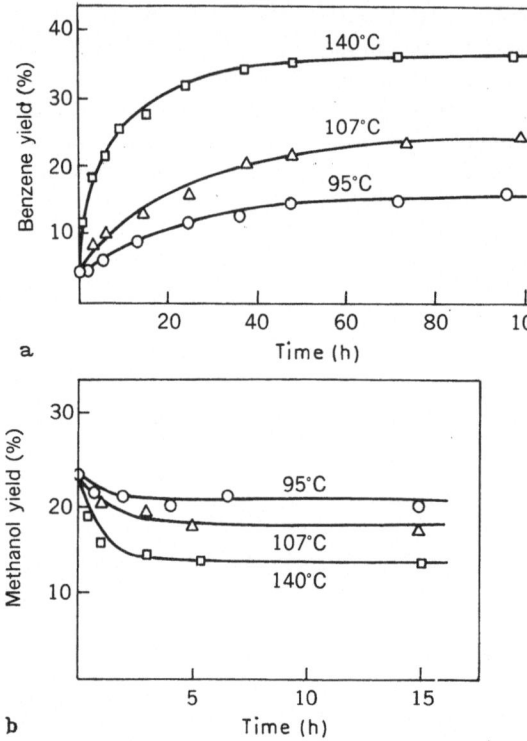

Fig. 3.3.5a, b. Yields of ^{51}Cr recoil species as a function of time after dissolution of neutron-irradiated Cr(acac)$_3$ in benzene (from Ref. 3.3.33). **a** Fraction elutable with benzene (retention); **b** fraction elutable with methanol (ligand-deficient species)

postulated in irradiated solid Cr(acac)$_3$ [3.3.38]; in solutions after dissolution of the irradiated solids, they may likely be dissociated to give monomeric ^{60}Co- or ^{51}Cr-recoil species.

Tominaga and coworkers have studied the behavior of the ^{60}Co recoil species in benzene solutions of irradiated solid Co(acac)$_3$ in the presence of air or argon, and concluded that at least 50% of the ^{60}Co recoil species formed in the irradiated solid Co(acac)$_3$ consist of $^{60}Co(II)$ species including $^{60}Co(acac)_2$ [3.3.36]. The formation of $^{60}Co(II)$ species (i.e., $^{60}Co(acac)_2$) in irradiated Co^{III}(acac)$_3$ has also been demonstrated later in vacuum sublimation experiments (Fig. 2.2.10) [3.3.39].

Reactions of Decayed Atoms in the Solid Phase. We now examine recoil reactions following nuclear decay processes in the solid state. The recoil ^{80}Br atoms arising from isomeric transition in outer anions of ^{80m}Br-labeled cobaltammine- and rhodiumammine bromides [M(NH$_3$)$_5$X]Br$_{2,3}$ (M = Co, Rh; X = Cl, Br, I, NO$_2$, NCS) can substitute for a ligand group X and coordinate directly to the metal atom [3.3.40, 41]. The observed correlation of the ^{80}Br-for-X substitution yield with the stability of the parent complex ion suggests that the chemical environment may play an important role in determining the fate of decaying hot atoms. Such correlation has also been observed for energetic radiobromine atoms produced from (n, γ) and (n, 2n) reactions in the similar system [3.3.42, 43]. Yoshihara and coworkers have compared the results of $^{57}Ni(EC, \beta^+)^{57}Co$ reaction in ^{57}Ni-

labeled $[Ni(NH_3)_6]Cl_2$ with those of $^{58}Ni(\gamma, p)^{57}Co$ reaction in $[Ni(NH_3)_6]Cl_2$, and concluded that the bond disruption initiated by the EC-decay was more localized as compared with that caused by (γ, p) reaction [3.3.44]. The retention values in cobalt- and nickel tetraphenylporphine complexes have been compared for several nuclear processes including IT, EC, β^+, (γ, n) and (n, γ) [3.3.45, 46]. The retentions in the decay processes are nearly 80%, indicating that the anomalous Coulombic excitation initiated by highly converted IT process has been relaxed rapidly in solids; low retention values (4—8%) for (γ, n) and (n, γ) processes demonstrate that the bond rupture on the central metal is mainly caused by the mechanical factor rather than the electrostatic factor. By contrast, considerable disruption of bonds after the IT process has been observed in similar experiments with ^{58m}Co-labeled cobalt complexes such as $[Co(en)_2(NO_2)_2]NO_3$ [3.3].

Annealing and Solid State Exchange Reactions. One of the most important topics in solid phase hot atom chemistry is the annealing phenomenon, in which the recoil species are converted (mostly to the parent chemical forms) by application of heat, light, radiation, etc. Although theoretical approach to the kinetics of annealing behaviors has been proposed [3.3.48—50], very few systems have been actually analyzed with a reasonable success, probably due to lack of very accurate experimental data. At present we would rather promote solid state exchange studies with doped tracer ions to simulate the annealing reactions in the solid phase. After several earlier works with conventional techniques on transfer annealing in chromate, iodate and metal chelates [3.3.51—56], Nath and Klein have verified unambiguously the solid state exchange between doped $^{57}Co^{2+}$ and $[Co(bipy)_3](ClO_4)_3 \cdot 3H_2O$ by means of Mössbauer spectroscopy [3.3.57]. Similar exchange studies in the solid state have been carried out later on a number of metal complexes doped with trace metal ions either by classical or by Mössbauer technique. As is shown in Fig. 3.3.6, $^{103}Ru(acac)_3$ can be formed by heating

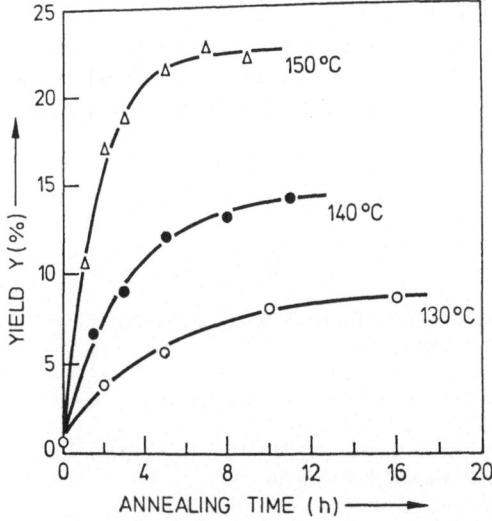

Fig. 3.3.6. Yield of $^{103}Ru(acac)_3$ on isothermal annealing of ^{103}Ru-doped $Co(acac)_3$ (taken from Ref. 3.3.58)

Co(acac)$_3$ doped with ^{103}Ru^{3+} [3.3.58]. In general, the mechanism of the solid state exchange of doped species and of the annealing of recoil species in the corresponding systems are postulated as essentially alike except for the radiation damage induced in the latter, because of the similarity in their kinetic behaviors. For instance, the ^{60}Co retention in neutron-irradiated Co(acac)$_3$ has been found to decrease quickly on dilution with Al(acac)$_3$, and to increase on dilution with Mn(acac)$_3$ [3.3.59]. In other experiments on Al(acac)$_3$ and Mn(acac)$_3$ doped with ^{57}Co^{2+}, the ^{57}Co^{2+} ions can readily exchange with the manganese complex, but not with the aluminum complex [3.3.60]. Hence, the dilution effect on the ^{60}Co-retention may be explained in terms of the solid phase exchange between the recoil ^{60}Co atoms and the host matrices.

Reactions in Mixed Systems

The results from the experiments on mixed crystals often provide useful information as to the size of the reactive zone produced by recoiling atoms in the solid state. Müller has studied the radiochemical yields of all ^{186}Re-labeled recoil species $[^{186}\text{ReCl}_n\text{Br}_{6-n}]^{2-}$ (n = 0 $\sim$ 6) following (n, γ) reaction in mixed systems K$_2$ReBr$_6$—K$_2$SnCl$_6$, K$_2$ReCl$_6$—K$_2$SnBr$_6$, and K$_2$ReBr$_6$—K$_2$OsCl$_6$ with various mixing ratios [3.3.61, 62]. It has been found that an appreciable fraction of the ligand atoms originally attached to rhenium atom tends to remain as ligands in the rhenium recoil species even in mixed crystals with maximum dilution. Based on these results, Müller has proposed 'disorder model' in which a small disordered zone (or about 5 Å in radius) is supposedly generated by recoil instead of a large 'hot zone' [3.3.1], and the fate of the recoiling atoms may be determined eventually by properties of the neighboring lattice. The assumption that the recoil generated reaction zone in solid is generally small and localized may be compatible with other experiments, e.g. very similar distribution of yields of the ^{60}Co recoil species observed for ^{59}Co(n, γ)^{60}Co reaction in both [Co(NH$_3$)$_6$] ·[Co(CN)$_6$] and [Co(NH$_3$)$_6$]·[Fe(CN)$_6$] [3.3.63] and similar product distribution and annealing behavior observed for the ^{58}Fe(n, γ)^{59}Fe reaction in both [^{58}Fe(bipy$_3$)$_3$ ·[^{56}Fe(CN)$_6$]$_2$ and [^{56}Fe(bipy)$_3$]$_3$·[^{58}Fe(CN)$_6$]$_2$ [3.3.64].

Ikeda and coworkers have studied the behavior of recoil ^{51}Cr in mixtures of [Co(NH$_3$)$_6$]Cl$_3$ and K$_2$CrO$_4$ powders and found that some of ^{51}Cr recoil atoms apparently leave the chromate particles and react in the cobaltammine particles, yielding [^{51}Cr(NH$_3$)$_6$]$^{3+}$, for example [3.3.65, 66].

References

[3.3.1] Harbottle, G., Sutin, N.: Advan. Inorg. Chem. Radiochem. *1*, 267 (1959)
[3.3.2] Harbottle, G.: Ann. Rev. Nucl. Sci. *15*, 89 (1965)
[3.3.3] Müller, H.: Angew. Chem. Int. Ed. Engl *6*, 133 (1967)
[3.3.4] Harbottle, G., Maddock, A. G., (eds.): *Chemical Effects of Nuclear Transformations in Inorganic Systems.* Amsterdam: North-Holland 1979
[3.3.5] Harbottle, G.: Hot Atom Chemistry Status Report, IAEA-PL-615/2 (1975)
[3.3.6] Yoshihara, K., Kudo, H.: J. Chem. Phys. *52*, 2950 (1970)
[3.3.7] Yoshihara, K., Mizusawa, T.: Radiochem. Radioanal. Lett. *9*, 263 (1972)

[3.3.8] Yoshihara, K., Fujita, A., Shiokawa, T.: J. Inorg. Nucl. Chem. *39*, 1733 (1977)

[3.3.9] Yoshihara, K., Yang, M. H.: Inorg. Nucl. Chem. Lett. *5*, 389 (1969)

[3.3.10] Aratani, M., Saito, N.: Shitsuryo Bunseki (Mass Spectrometry) *18*, 906 (1970)

[3.3.11] Ingalls, R., De Pasquali, G.: Phys. Lett. *15*, 262 (1965)

[3.3.12] Friedt, J. M., Adloff, J. P.: Compt. Rend. *264C*, 1356 (1967)

[3.3.13] Sano, H., Iwagami, H.: Chem. Commun. *1971*, 1637

[3.3.14] Sano, H., Sato, K., Iwagami, H.: Bull. Chem. Soc. Jpn. *44*, 726 (1971)

[3.3.15] Sano, H.: Radioisotopes *24*, 357 (1975)

[3.3.16] Sano, H., Harada, M., Endo, K.: Bull. Chem. Soc. Jpn. *51*, 2583 (1978)

[3.3.17] Gallagher, P. K.: in *Applications of Mössbauer Spectroscopy, Vol. 1* R. L. Cohen (ed.). New York: Academic Press 1976.

[3.3.18] Gütlich, P., Link, R., Trautwein, A.: *Mössbauer Spectroscopy and Transition Metal Chemistry*, Inorganic Chemistry Concepts Vol. 3. Berlin, Heidelberg, New York: Springer 1978

[3.3.19] Saito, N., Sano, H., Tominaga, T., Ambe, F.: Bull. Chem. Soc. Jpn. *38*, 681 (1965)

[3.3.20] Sugimori, A., Tsuchihashi, G.: Bull. Chem. Soc. Jpn. *34*, 449 (1961)

[3.3.21] Sano, H.: Radioisotopes *24*, 357 (1975)

[3.3.22] Sano, H., Sato, K., Iwagami, H.: Bull. Chem. Soc. Jpn. *44*, 2570 (1971)

[3.3.23] Gal, J., Hadari, Z., Yanir, E.: J. Inorg. Nucl. Chem. *32*, 2509 (1970)

[3.3.24] Stone, J. A., Phillinger, W. L.: Phys. Rev. Lett. *13*, 200 (1964)

[3.3.25] Rauscher, H., Harbottle, G.: J. Inorg. Nucl. Chem. *4*, 155 (1957)

[3.3.26] Müller, H.: Angew. Chem. Int. Ed. Engl., *6*, 133 (1967)

[3.3.27] Stucky, G. L., Kiser, R. W.: Radiochim. Acta, *11*, 5 (1967)

[3.3.28] Saito, N., Tominaga, T., Sano, H.: Bull. Chem. Soc. Jpn. *35*, 365 (1962); *36*, 232 (1963); *38*, 1407 (1965)

[3.3.29] Saito, N., Sano, H., Tominaga, T.: Chem. Ind. (London) *1964*, 1622

[3.3.30] Tominaga, T., Sakai, T.: Bull. Chem. Soc. Jpn. *45*, 1237 (1972)

[3.3.31] Tominaga, T., Sakai, T., Fujiwara, K.: Bull. Chem. Soc. Jpn. *44*, 3036 (1971)

[3.3.32] Tominaga, T., Nishi, Y.: Radiochem. Radioanal. Lett. *8*, 151 (1971)

[3.3.33] Gainar, I., Ponta, A.: Radiochem. Radioanal. Lett. *7*, 79 (1971)

[3.3.34] Nishi, Y., Tominaga, T.: Radioisotopes *23*, 700 (1974)

[3.3.35] Sarup, S., Nath, A.: J. Inorg. Nucl. Chem. *29*, 299 (1967)

[3.3.36] Tominaga, T., Nishi, Y., Motohashi, E.: Radiochem. Radioanal. Lett. *18*, 15 (1974)

[3.3.37] Venkateswarlu, K. S., Kishore, K.: Radiochim. Acta *15*, 70 (1971)

[3.3.38] Omori, T., Yeh, Yu-Cahi, Shiokawa, T.: Radiochem. Acta *18*, 1 (1972)

[3.3.39] Amano, R., Sakanoue, M.: Radiochem. Radioanal. Lett. *19*, 197 (1974)

[3.3.40] Saito, N., Ito. S., Tominaga, T.: Bull. Chem. Soc. Jpn. *38*, 504 (1965)

[3.3.41] Schmidt, G. B., Herr, W.: Z. Natur. Forsch. *18a*, 505 (1963)

[3.3.42] Saito, N., Tominaga, T., Sano, H.: Bull. Chem. Soc. Jpn. *33*, 120 (1960); ibid. *35*, 63 (1962)

[3.3.43] Saito, N., Takeda, M., Tominaga, T.: Bull. Chem. Soc. Jpn. *40*, 690 (1967)

[3.3.44] Yoshihara, K.: Nature *204*, 1296 (1964)

[3.3.45] Ikeda, N., Shoji, H., Sakai, Y., Nakajima, S.: Kakuriken Kenkyu Hokoku (Tohoku Univ.) *9*, 253 (1976)

[3.3.46] Ikeda, N., Shoji, H., Sakai, Y., Nakajima, S.: Kakuriken Kenkyu Hokoku (Tohoku Univ.) *9*, 177 (1976)

[3.3.47] Lazzarini, E., Lazzarini, A. L. F.: J. Inorg. Nucl. Chem. *39*, 207 (1977)

[3.3.48] Vand, V.: Proc. Phys. Soc. London *55*, 222 (1943)

[3.3.49] Primak, W.: Phys. Rev. *100*, 1677 (1955)

[3.3.50] Yaffe, L. (ed.): *Nuclear Chemistry*, Vol. 2. New York: Academic Press 1968

[3.3.51] Kaucic, S., Vlatkovic, M.: Croat. Chem. Acta *35*, 305 (1963)

[3.3.52] Apers, D. J., Collins, K. E., Collins, C. H., Ghoos, Y. F., Capron, P. C.: Radiochem. Acta *3*, 18 (1964)

[3.3.53] Collins, C. H., Collins, K. E., Ghoos, Y. F., Apers, D. J.: Radiochim. Acta *4*, 211 (1965)

[3.3.54] Nath, A., Khorana, S., Mathur, P. K., Sarup, S.: Indian J. Chem. *4*, 51 (1966)
[3.3.55] Nath, A., Khorana, S.: J. Chem. Phys. *46*, 2858 (1967)
[3.3.56] Khorana, S., Nath, A.: J. Phys. Chem. Solids *28*. 1081 (1967)
[3.3.57] Nath, A., Klein, M. P.: Nature *224*, 794 (1969)
[3.3.58] Meinhold, H., Reichold, P.: Inorg. Nucl. Chem. Lett. *6*, 253 (1970)
[3.3.59] Shankar, J., Venkateswarlu, K. S., Lal, M.: Radiochim. Acta *4*, 52 (1965)
[3.3.60] Ramshesh, V.: J. Inorg. Nucl. Chem. *31*, 3878 (1969)
[3.3.61] Müller, H.: Naturwiss. *49*, 182 (1962)
[3.3.62] Müller, H.: J. Inorg. Nucl. Chem. *27*, 1745 (1965)
[3.3.63] Saito, N., Tominaga, T., Sano, H.: Nature *194*, 466 (1962)
[3.3.64] Siekierska, K. E., Fenger, J., Maddock, A. G.: J. Chem. Soc. (A) *1973*, 1086
[3.3.65] Ikeda, N., Saito, K., Tsuji, K.: Radiochim. Acta *13*, 90 (1970)
[3.3.66] Ikeda, N., Kujirai, O.: Radioisotopes *20*, 56 (1971)

4 Applications of Hot Atom Chemistry and Related Topics

4.1 Applications in Inorganic, Analytical and Geochemistry

4.1.1 Applications in Inorganic and Analytical Chemistry

Hot atom applications in inorganic chemistry mainly consist in the production of enriched radioisotopes and synthesis of labeled compounds. Samples of radioisotopes in high specific activities have often been obtained by separating the fraction enriched in desired isotopes having recoiled from the parent systems. A number of labeled inorganic and organometallic compounds have been prepared by means of recoil synthesis through various types of nuclear reactions. Energetic radionuclides arising from the nuclear fission, or radioactive tracer ions accelerated in a chemical accelerator can be implanted into solid matrices to produce labeled compounds or to introduce 'impurities' into the target materials. An important contribution of the nuclear technique to inorganic preparative chemistry is that unknown compounds unavailable by ordinary chemical synthesis have been obtained as products from the beta decay.

Hot atom chemistry has also found applications in analytical chemistry. It is well known that unknown nuclides produced in the nuclear process have been separated by the recoil technique in the experiment to confirm formation of a new element, mendelevium [4.1.1]. Hot atom chemists may also be interested in topics such as molecular activation analysis, and loss of recoiled atoms in activation analysis, the significance of which is still open to further investigation.

Production of Enriched Radioisotopes. Enrichment of radioisotopes was one of the most practical applications of hot-atom chemistry in its early days. The technique for radioisotope enrichment through recoil reactions has been known as the Szilard-Chalmers process, named after the scientists who first discovered the hot-atom reactions of radioiodine in neutron-irradiated ethyl iodide [4.1.2]. Since the recoiled radioactive atoms are inevitably accompanied by non-active isotopic atoms arising from radiolysis or pyrolysis of the target material during irradiation in the nuclear reactor, the enrichment factor (i.e. the ratio of specific activity of the separated fraction to specific activity of the overall sample before separation) attainable with this technique is limited by stabilities of the target material towards concomitant radiation or heat in the reactor. Although radioisotopes in high specific activities can now be prepared simply by using reactors with higher neutron fluxes ($\sim 10^{15}$ n cm$^{-2} \cdot$s^{-1}), certain enriched radioisotopes (e.g. ^{51}Cr, ^{64}Cu, ^{32}P) are still produced via recoil process in medium-size reactors with neutron fluxes of $10^{12} - 10^{13}$ n $cm^{-2} \cdot s^{-1}$. For example, highly enriched $^{51}Cr^{3+}$ samples can be obtained by the precipitation method from neutron-irradiated potassium chromate.

Recoil Synthesis of Inorganic Compounds. Hot-atom reactions have been used to prepare labeled compounds and such an application is often called 'recoil synthesis'. Recoil synthesis of organic molecules will be discussed in Sect. 4.3. A number of labeled transition metal complexes and organometallic compounds have been synthesized through recoil processes following various types of nuclear reactions, e.g. (n, γ), (n, 2n), (γ, n), or isomeric transition. In Table 4.1.1 are summarized typical recoil synthesis reactions, i.e. recoil labeling of the ligands and the central atom of metal complexes [4.1.3—18]. As has been discussed in Sect. 3.3, the yield for labeling in ligands (i.e. ligand yield) of metal complexes depends on the chemical properties of the complexes, e.g. stability of bonding between the metal and the ligand to be substituted for [4.1.3—7]. A variety of recoil products are generally produced in labeling of the central atom. The relative yields of these labeled products also depend on the chemical properties of the parent systems, and can be explained by assuming a small reactive zone in which former ligands and outer anions compete to recombine with the 'labeled' central atom [4.1.10—15]. From the preparative point of view, we may choose the most adequate system and irradiation-annealing conditions to obtain a particular labeled product in a reasonable yield, on the basis of the systematic knowledge of the hot-atom reactions accumulated to date.

In view of the enormous kinetic energy and the large recoil range, the fission recoil synthesis should be different from labeling processes using the other nuclear transformations: fission recoil atoms can penetrate deeply into the solid matrix and react within it, even when the fission fragments have been produced in a heterogeneous mixture with the target material. A typical recoil labeling with fission fragments consists in the displacement of the central metal atom in an organometallic molecule such as metallocenes, metal carbonyls, and metal acetylacetonates [4.1.19—23]. Baumgärtner and coworkers have irradiated with thermal neutrons mixtures of powders of U_3O_8 and ferrocene or chromium hexacarbonyls, and separated $^{103}Ru\text{-}(C_5H_5)_2$ or $^{99}Mo(CO)_6$ in 40—60% yields by sublimation. The latter system was used for rapid separation by sublimation of the short-lived isotope ^{103}Mo (60s) as $^{103}Mo(CO)_6$ [4.1.24].

The fission recoil synthesis may be thus regarded as a process for implantation of radioactive species into the target material utilizing their own recoil energies. In ion-implantation studies, however, radioactive ions accelerated artificially are injected into the solid matrix, and the chemical states and environment of the implanted atoms can be investigated either by in-situ physical techniques (e.g. Mössbauer spectroscopy) or by conventional chemical methods based on etching or sectioning of the 'implanted surface'. Ion implantation experiments have been attempted by hot atom chemists in complex solids and simple systems (e.g. ^{32}P or ^{35}S-implanted alkali chloride single crystals) so as to compare the behaviors of implanted and nuclear-recoil-generated atoms [4.1.25—28]. Unlike solid-phase hot-atom chemistry, the ion implantation technique provides de-

Table 4.1.1. Typical recoil synthesis reactions in solid metal complexes

Types of labeling reactions	Systems	Nuclear reactions	Typical reactions and products	Ref.
Labeling in ligands	Ammine complexes	(n, γ) (n, 2n) (d, n) (t, n) IT	$[M(NH_3)_5X]Br_{2,3} \xrightarrow{Br^*} [M(NH_3)_5Br^*]^{2+}$	4.1.3—7
			$[M(en)_2X_2]Br_{1,3} \xrightarrow{Br^*} [M(en)_2XBr^*]^{1,2+}$ (M = Co, Rh; X = F, Cl, Br, I, NH₃, NCS, NO₂, H₂O, etc.)	4.1.3—5
			$[Co(NH_3)_6]Cl_3 \xrightarrow{Cl^*} [Co(NH_3)_5Cl^*]^{2+}$	4.1.8
			$[Co(NH_3)_6]F_3 \xrightarrow{F^*} [Co(NH_3)_5F^*]^{2+}$	4.1.9
			$[Co(NH_3)_6](NO_3)_3 \xrightarrow{F^*} [Co(NH_3)_5F^*]^{2+}$	4.1.9
Labeling in central metal atoms	Ammine complexes Nitroammine complexes Cyano complexes	(n, γ) (n, 2n) (γ, n) IT	Synthesis type reactions:	
			$[Co(NH_3)_6X_3] \xrightarrow{Co^*} [Co^*(NH_3)_5X]^{2+}$, $[Co^*(NH_3)_4X_2]^+$, $[Co^*(NH_3)_3X_3]$, ... (X = F, Cl, Br, I, NO₂, NO₃, etc.)	4.1.10—14
			$[Co(NH_3)_5NO_2] \xrightarrow{Co^*} [Co^*(NH_3)_4(NO_2)_2]^+$, $[Co^*(NH_3)_3(NO_2)_3]$, $[Co^*(NH_3)_2(NO_2)_4]^-$, ...	4.1.14, 15
			Central atom substitution:	
			$[Co(NH_3)_6][Fe(CN)_6] \xrightarrow{Co^*} [Co^*(CN)_6]^{3-}$	4.1.16
			$[Co(NH_3)_2(CrO_4)_3 \xrightarrow{Cr^*} [Cr^*(NH_3)_6]^{3+}$	4.1.17
			Isomerization:	
			cis-$[Co(NH_3)_4(NO_2)_2]^+$ $\xrightarrow{Co^*}$ trans-$[Co^*(NH_3)_4(NO_2)_2]^+$	4.1.18

tailed information regarding the effects on the behavior of implanted species, of the incident beam energy and direction with respect to the crystal orientation (e.g. channeling phenomenon). Formation of $^{56}Mn(CO)_5$ and $^{56}Mn(CO)_4$ in small yields has been reported on implantation of $^{56}Mn^+$ ions into $Cr(CO)_6$ matrix [4.1.29]. Inorganic chemists may be interested in whether or not a mixed manganese-chromium carbonyl can be formed through the interaction of these manganese species with the host matrix.

Since the chemical effects following pure beta decay or beta-decay accompanied by poorly converted γ-transitions may not be too drastic, molecules can often survive such nuclear transformations without complete disruption of bonds. Hence, molecules containing the daughter atom in place of the parent have been synthesized through beta decay, some of which are unknown compounds never obtained in ordinary preparative chemistry. The perbromate ion, which was missing among the perchlorate and periodate series, has been synthesized via beta decay of ^{83}Se in selenate [4.1.30]:

$$^{83}SeO_4^{2-} \xrightarrow{\beta^-} {}^{83}BrO_4^-.$$

The high oxidation state ($+7$) has been attained by an oxidizing process in the beta decay, or increment of a positive charge on the daughter atom. In the beta decay of the central atom of organometalic compounds are similarly obtained new products [4.1.31]:

$$^{105}Ru(C_5H_5)_2 \xrightarrow{\beta^-} {}^{105}Rh(C_5H_5)_2^+ \xrightarrow{+e^-} {}^{105}Rh(C_5H_5)_2.$$

Xenon compounds have also been prepared by way of beta decay of the corresponding ^{129}I-labeled iodine compounds [4.1.32—34]:

$$K^{129}ICl_4 \xrightarrow{\beta^-} {}^{129}XeCl_4.$$

The unknown xenon compounds (oxide and halides) thus radiosynthesized have been identified by their Mössbauer spectra (Fig. 4.1.1).

Analytical Applications. Recoil atoms can be separated from the parent material according either to their positive charges or to large kinetic energies. While charged recoil species are collected electrostatically on a catcher foil, fission recoil atoms with enormous energies can be separated into powders of a catcher compound mixed with the fissile material (e.g. U_3O_8-graphite mixtures). This may provide a unique means of separation in certain analytical applications. The information regarding the chemical identity of recoiled atoms may sometimes be important in activation analysis. It is probable that the activated atoms may escape from the surface of the sample because of the excess recoil energy, or that they may eventually assume 'volatile' chemical forms after the recoil reactions. Such a problem has been actually encountered in activation analysis of mercury or halogens.

Although activation analysis is a means for elemental analysis, there have been a few attempts to develop molecular activation analysis, a technique to obtain information regarding the chemical identity of the compound containing a particular element [4.1.35—38]. It is essential for this technique that the 'retention' of the element in the non-exchanging compound under study should

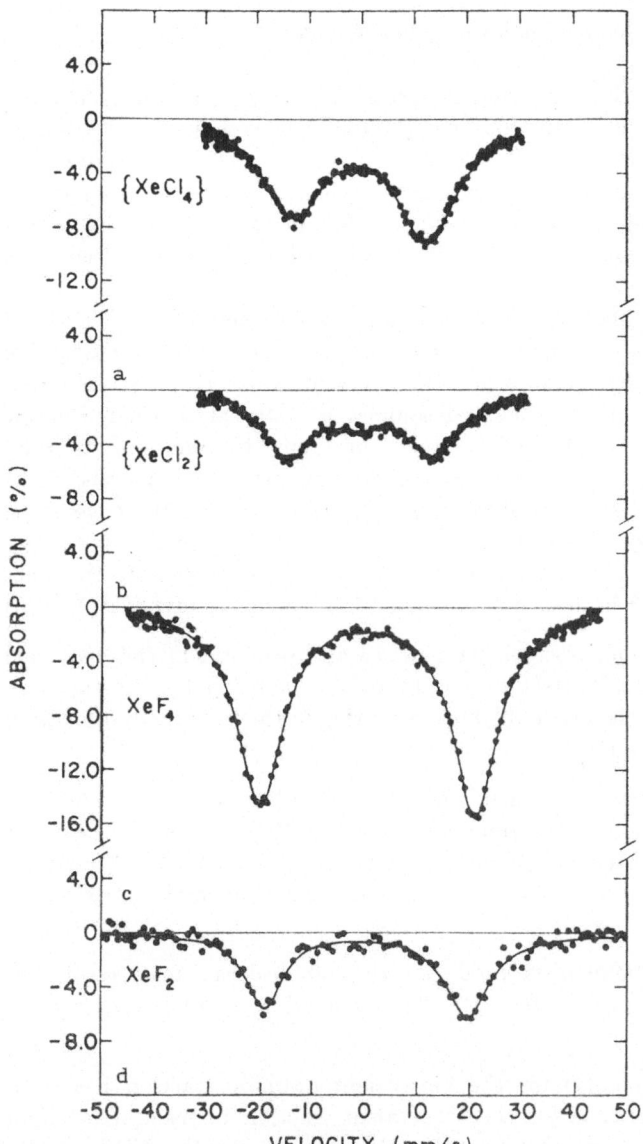

Fig. 4.1.1. Mössbauer spectra of xenon halides. Mössbauer emission spectra of (a) $K^{129}ICl_4 \cdot$ H_2O and (b) $K^{129}ICl_2 \cdot H_2O$ vs. a standard absorber; Mössbauer absorption spectra of (c) XeF_4 and (d) XeF_2 (taken from Ref. 4.1.33)

remain reasonably constant for the analytical conditions employed. Preliminary experiments to determine mercury in methylmercury compounds [4.1.35—37] and chlorine in chloranilic acid [4.1.38] have shown that the concept of molecular activation analysis may be valid at least for certain controlled systems. Whether this technique can be generally successful in actual analytical applications still awaits more detailed investigation.

4.1.2 Applications and Related Topics in Geochemistry

Recoil phenomena sometimes play an important role in geochemistry. The isotopic fractionations of naturally occurring radionuclides have been reported in various geochemical systems. The formation of fission tracks in radioactive minerals is also based on recoil and radiation effects by energetic heavy fission fragments. However, we are not discussing on fission track studies in further detail since they are mainly related to geochronological problems rather than hot atom chemistry.

It has been generally believed that the ratio of radioactivities $^{234}U/^{238}U$ in natural geochemical systems should be unity in accordance with a radioactive equilibrium until the Cherdyntsev's first discovery of radioactive disequilibrium between ^{234}U and ^{238}U in some natural samples [4.1.39]. Since then, isotopic fractionations in natural radionuclides (e.g., uranium, thorium and radium) have been investigated in a number of geochemical systems (minerals, rocks and ground water, etc.) by means of isotopic analysis of such samples or leaching experiments [4.1.40—44]. The isotope effects are generally negligible for heavy elements, and the observed isotopic fractionations have been ascribed to the following mechanisms in which hot-atom chemistry is more or less involved [4.1.45]:

a) The daughter nuclides formed via α-decay are displaced in the disturbed crystal lattice by the recoil effect and stabilized on more labile sites than the nuclides incorporated in the original lattice, resulting in the isotopic fractionation in the leaching process [4.1.40].

b) The parent nuclides and their isotopic descendent nuclides sometimes assume different oxidation states and thus indicate different chemical behavior because of the consequences of nuclear transformations or because of the migration of the latter nuclides by recoil onto the sites more vulnerable to oxidation (selective exposure) [4.1.41].

c) The daughter radionuclides produced through α-decay may recoil out from the solid surface and ejected into the surrounding system, e.g., ground water [4.1.43].

The $^{234}U/^{238}U$ activity ratio for the hexavalent uranium fraction has been found to be larger than that for the tetravalent fraction in natural uranium oxides, pitchblende and uranium black, and marine phosphorite: the ^{234}U generally tends to concentrate in the hexavalent state in minerals [4.1.41, 42, 44]. In another uranium ore, euxenite, however, ^{234}U has been found to concentrate in the tetravalent state [4.1.46]. When the euxenite sample was subjected to a leaching process, the $^{234}U/^{238}U$ ratio in the leachates has proved to be smaller than unity, probably due to the preferred leaching of hexavalent uranium [4.1.46]. The activity ratios $^{228}Th/^{232}Th$ and $^{224}Ra/^{228}Ra$ also measured in the leachates from the euxenite sample turn out to be larger than unity, demonstrating that the radiogenic nuclides arising from α-decay are leached out preferentially [4.1.46]. The behaviors of such radioactive nuclides on leaching of the minerals can be thus generally understood in terms of their fate in the disturbed lattice, or the nature of the damage produced by the recoil effect (size and deposited energy, annealing, etc.).

An interesting experiment has been reported on the application of the hot-atom technique in geochemistry—application of neutron activation analysis to locate radiogenic xenon isotopes in meteorite [4.1.47]. The meteorite sample contains two gaseous xenon components which are released at different temperatures on heating: They are (i) adsorbed xenon gas with the isotopic abundances similar to those of atmospheric xenon, and (ii) ^{129}Xe-enriched xenon gas, possibly produced via β-decay of an extinct isotope, ^{129}I, incorporated in meteorite. On neutron irradiation of the meteorite sample, ^{128}I is produced through (n, γ) reaction on ^{127}I in the iodide-containing phase, and then decays to ^{128}Xe. When the irradiated meteorite was heated progressively, the ^{129}Xe-enriched xenon gas and ^{127}I(n, γ)-produced ^{128}Xe were released simultaneously. Since ^{128}I and ^{128}Xe cannot recoil out of the original microcrystals after (n, γ) and β-decay recoil, they are still likely located within the iodide-containing phase. Hence, the simultaneous release of ^{129}Xe and ^{128}Xe has confirmed the assumption that the ^{129}Xe atoms are located on the iodide sites in the meteorite.

Recently great concern has been expressed for the stratospheric ozone depletion by chlorofluorocarbons emitted into the atmosphere [4.1.48]. These anthropogenic chlorinated molecules gradually reach the stratosphere and are decomposed by u.v. light to release atomic chlorine which in turn destroys ozone through the ClO_x catalytic chain reaction. Short-lived radioisotopes ^{38}Cl (37 min) and ^{34}Cl (55 min) produced by cosmic ray bombardment of atmospheric ^{40}Ar also participate in the chlorine cycle and form labeled compounds such as HCl*, Cl*O, Cl*ONO$_2$, etc. Measurements of these species in air samples are useful in understanding of the chemistry of such compounds in the atmosphere [4.1.49].

In concluding this section, we may point out the possibility that hot atom chemists will become in near future more interested in what is taking place in the atmosphere of other planets.

References

[4.1.1]   Ghiorso, A., Harvey, B. G., Choppin, G. R., Thompson, S. G., Seaborg, G. T.: Phys. Rev. *99*, 1518 (1955)
[4.1.2]   Szilard, L., Chalmers, T. A.: Nature (London) *134*, 462 (1934)
[4.1.3]   Saito, N., Tominaga, T., Sano, H.: Bull. Chem. Soc. Jpn. *35*, 63 (1962)
[4.1.4]   Saito, N., Itoh, S., Tominaga, T.: Bull. Chem. Soc. Jpn. *38*, 504 (1965)
[4.1.5]   Saito, N., Takeda, M., Tominaga, T.: Bull. Chem. Soc. Jpn. *40*, 690 (1967)
[4.1.6]   Tominaga, T., Ishii, M., Saito, N.: Radioisotopes *20*, 579 (1971)
[4.1.7]   Schmidt, G. B., Herr, W.: Z. Naturforsch. *18a*, 505 (1963)
[4.1.8]   Ikeda, N., Yoshihara, K., Yamagishi, S.: Bull. Chem. Soc. Jpn. *34*, 140 (1961); Radiochim. Acta *3*, 13 (1964)
[4.1.9]   Saito, N., Ambe, F., Ambe, S., Shimamura, A.: Bull. Chem. Soc. Jpn. *43*, 284 (1970)
[4.1.10] Harbottle, G.: Chem. Effects. Nucl. Transform., Proc. Symp. Prague *1960*, 301
[4.1.11] Saito, N., Tominaga, T., Sano, H.: Bull. Chem. Soc. Jpn. *35*, 365 (1962)
[4.1.12] Saito, N., Tominaga, T., Sano, H.: Bull. Chem. Soc. Jpn. *38*, 1407 (1965)
[4.1.13] Saito, N., Tominaga, T., Sano, H.: Bull. Chem. Soc. Jpn. *36*, 230 (1963)
[4.1.14] Saito, N., Tominaga, T, Sano, H.: Bull. Chem. Soc. Jpn. *33*, 1921 (1960); J. Inorg. Nucl. Chem. *24*, 1539 (1962)

[4.1.15] Saito, N., Tominaga, T., Sano, H.: Bull. Chem. Soc. Jpn. *36*, 232 (1963)
[4.1.16] Saito, N., Tominaga, T., Sano, H.: Nature (London) *194*, 466 (1962)
[4.1.17] Ikeda, N., Saito, K., Tsuji, K.: Radiochim. Acta *13*, 90 (1970)
[4.1.18] Saito, N., Sano, H., Tominaga, T., Ambe, F., Fujino, T.: Bull. Chem. Soc. Jpn. *35*, 744 (1962)
[4.1.19] Baumgärtner, F., Reichold, P.: Z. Naturforsch. *16a*, 945 (1961)
[4.1.20] Kienle, P., Weckermann, B., Baumgärtner, F., Zahn, U.: Naturwiss. *49*, 295 (1962)
[4.1.21] Baumgärtner, F., Reichold, P.: Z. Naturforsch. *16a*, 374 (1961)
[4.1.22] Baumgärtner, F., Schoen, A.: Radiochim. Acta *3*, 141 (1964)
[4.1.23] Meinhold, H., Reichold, P.: Radiochim. Acta *11*, 175 (1969)
[4.1.24] Kienle, P., Baumgärtner, F., Zahn, U.: Radiochim. Acta *1*, 84 (1963)
[4.1.25] Anderson, T. A., Langvad, T., Sorensen, G.: Nature (London) *218*, 1158 (1968)
[4.1.26] Wolf, G. K., Fritsch, T.: Radiochim. Acta *11*, 194 (1969)
[4.1.27] Anderson, T. A., Ebbensen, A.: Trans. Faraday Soc. *67*, 3540 (1971)
[4.1.28] Kasrai, M., Maddock, A. G., Freeman J. H.: Trans. Faraday Soc. *67*, 2108 (1971)
[4.1.29] Jenkins, G. M., Wiles, D. R.: Chem. Commun. *1972*, 1177
[4.1.39] Appelman, E. H.: J. Am. Chem. Soc. *90*, 1900 (1968)
[4.1.31] Baumgärtner, F.: Chem. Effects Nucl. Transform. Proc. Symp. Vienna, *1965*, Vol. II, p. 507
[4.1.32] Perlow, G. J., Perlow, M. R.: Chem. Effects Nucl. Transform. Proc. Symp. Vienna, *1965*, Vol. II, p. 443
[4.1.33] Perlow, G. J., Perlow, M. R.: J. Chem. Phys. *48*, 955 (1968)
[4.1.34] Perlow, G. J., Yoshida, H.: J. Chem. Phys. *49*, 1474 (1968)
[4.1.35] Heitz, C.: Bull. Soc. Chim. France, *1967*, 2442
[4.1.36] Wheeler, O. H., McClin, M. L.: Intern. J. Appl. Radiat. Isotopes *18*, 788 (1967)
[4.1.37] DeJong, I. G., Omori, T., Wiles, D. R.: Chem. Commun., *1974*, 189
[4.1.38] Grant, P. M., Rowland, F. S.: Hot Atom Chemistry Status Report IAEA-PL-615/13 (1975)
[4.1.39] Cherdyntsev, V. V.: in *Tr. III Sessii Komissi po Opredelniyu Absolyutnogo Vozrasta Geologicheskikh Formatsii.* Moscow: Akad. Nauk SSSR 1965
[4.1.40] Rosholt, J. N., Shields, W. R., Garner, E. L.: Science *13*, 224 (1963)
[4.1.41] Chalov, P. I., Merkulova, K. I.: Dokl. Akad. Nauk SSSR, *167*, 146 (1966)
[4.1.42] Chalov, P. I., Merkulova, K. I.: Geokhimiya, 872 (1968)
[4.1.43] Kigoshi, K.: Science *173*, 47 (1971)
[4.1.44] Klody, Y., Kaplan, I. R.: Geochim. Cosmochim. Acta *34*, 3 (1970)
[4.1.45] Starik, I. E.: *Osnovy Radiokhimii.* Moscow: Akad. Nauk SSSR 1960
[4.1.46] Kobashi, A., Sato, J., Saito, N.: Radiochim. Acta *26*, 107 (1979)
[4.1.47] Reynolds, J. H.: Ann. Rev. Nucl. Sci. *17*, 253 (1967)
[4.1.48] Molina, M. J., Rowland, F. S.: Nature *249*, 810 (1974); F. S. Rowland, M. J. Molina: Rev. Geophys. Space Phys., *13*, 1 (1975)
[4.1.49] Rowland, F. S.: Geophys. Res. Lett. *5*, 9 (1978)

4.2 Applications in Physical Chemistry

As has been clarified in the previous chapters, the kinetic analysis of the reactions of energetic atoms arising from nuclear recoil can often be accomplished with the knowledge obtained from conventional experimental techniques. In some cases, however, the recoil studies are employed to furnish information regarding various aspects of gaseous reactions without any special difficulties. Unimolecular decomposition modes of various excited molecules, reactions of carbene types of intermediates, and ion-molecule reactions of various cations, formed directly or indirectly in the course of reactions of recoil atoms, are some of such examples [4.2.1].

Furthermore, under certain selected and carefully designed experimental conditions, the recoil methods have proved to be useful for determination of kinetic parameters. The success in such applications directly leads not only to expansion of the field of hot atom chemistry but also to further inspection of the reaction processes involved. Typical cases will be described briefly in the following sections.

Determination of Kinetic Parameters

For the recoil T reactions with $c\text{-}C_6H_{10}$, Fee et al. [4.2.2] have estimated apparent rate constant and s parameter in RRK-treatment of unimolecular decomposition of $c\text{-}C_6H_9T$ as $5.1 \times 10^6 \text{ s}^{-1}$ and 24, respectively. Furthermore, the scavenger dependence of yields with an obvious radical precursor was used to determine the relative rate constants of abstraction (k_1) vs addition (k_2) of that radical in the alkene parent compound:

ex. $CH_2T + $ ⬡ $\xrightarrow{k_1} CH_3T + R$

$CH_2T + $ ⬡ $\xrightarrow{k_2}$ ⬡$-CH_2T$

The results are summarized in Table 4.2.1.

Table 4.2.1. Radical reaction rate constants at 25 °C (taken from Ref. 4.2.2; Copyright 1974, American Chemical Society)

Reactants		k_2,[a] 10^6 cm³ mol⁻¹ s⁻¹	k_1/k_2,[b] lit.	k_1/k_2, this work
Radical	Double bond			
Methyl	O_2	300,000	na[c]	—[d]
	SO_2	5,000	na	—[d]
	H_2S	3,000[e]	na	—[d]
	Butadiene	160	—[d]	0.0019[f]
	Ethylene	1.2	0.015	0.0028
	Propylene	1.2	0.096	0.060
	1-Butene	1.0	0.37	0.075
	Isobutene	4.0	0.040	0.086
	Cyclohexene	—[d]	—[d]	0.36
Ethyl	Ethylene	—[d]	—[d]	0.091
	Cyclohexene	—[d]	—[d]	0.37
Propyl	Propylene	—[d]	—[d]	0.15
n-Butyl	1-Butene	—[d]	—[d]	0.16
Ethyl	Cyclohexene	—[d]	—[d]	0.35
Cyclohexyl	Cyclohexene	—[d]	—[d]	0.29
n-Hexenyl	Cyclohexene	—[d]	—[d]	0.32

[a] k_2 = rate constant of radical addition to the double bond.
[b] k_1/k_2 = ratio of abstraction/addition rate constants of radicals with alkenes.
[c] Not applicable.
[d] Not determined.
[e] For H-atom abstraction.
[f] The typical sample contained 110 Torr of parent hydrocarbon (55 Torr in the case of cyclohexene), 16 Torr of ³He and ~ 10 mol % scavenger when used.

The energetic ^{18}F atoms formed by recoil methods are well suited for various studies on fluorine atom reactions without such experimental difficulties as associated with the macroscopic amounts of F, HF, F_2 [4.2.3]: for example, glass containers show no visible etching, and heat increase due to exothermicity of the reactions is always negligible. The systems always contain large excess of an inert diluent such as SF_6 which serves both as the source of ^{18}F and as the moderator for the excess recoil energy of [18]F. Under these conditions, rates of hydrogen abstraction by ^{18}F relative to the rate of addition to acetylene have been measured in H_2, D_2, CH_4, and others [4.2.4]. The measurements have been based on competitive diminution in the yield of $CH_2=CH^{18}F$ vs RH (or RD)/C_2H_2, in RH (or RD)$-CH\equiv CH-HI-SF_6$ systems. It is also found [3] that the initial ^{18}F addition step is approximately equal in rates for C_2H_2 and $CHCl=CHCl$ but six times slower for $CFCl=CFCl$.

Root and coworkers [4.2.5] have applied steady state considerations to the reactions of ^{18}F in C_3F_6-moderated systems. In their reaction systems, the reactions involved are:

$$^{18}F + H_2 \rightarrow H^{18}F + H, \qquad\qquad (4.2.3a)$$

$$^{18}F + D_2 \rightarrow D^{18}F + D, \qquad\qquad (4.2.3b)$$

$$^{18}F + C_3F_6 \rightarrow \text{labeled products.} \qquad\qquad (4.2.4)$$

By employing ordinary linear differential rate equations, Eq. (4.2.5) is derived:

$$(k_4/k_3) = (1 - X)\,(1 - Y_{H^{18}F})/(X \cdot Y_{H^{18}F}) \qquad\qquad (4.2.5)$$

where X is the C_3F_6 mole fraction in each sample, $Y_{H^{18}F}$ the yield of $H^{18}F$. Figure 4.2.1 demonstrates that the experimental composition dependence calculated from Eq. (4.2.5) is linear within the experimental accuracy so that linear regression analysis has been used to systematize these results. The regression equations are given in Table 4.2.2 together with the estimated rate constant values extrapolated to the extreme mixture composition. These latter values reflect the relative reaction rates for Eqs. (4.2.3) vs (4.2.4) in the hypothetical pure H_2 (or D_2) and C_3F_6 reaction systems.

Table 4.2.2. Composition dependent relative rate constants for unmoderated H_2/C_3F_6 and D_2/C_3F_6 systems[a] (taken from Ref. 4.2.5)

T ($^\circ$K)	b(T)	m(T)	K(0, T)	K(1, T)
H_2 system				
273	1.024 ± 0.04	-0.250 ± 0.08	1.024 ± 0.04	0.774 ± 0.09
303	0.827 ± 0.03	-0.269 ± 0.03	0.827 ± 0.03	0.558 ± 0.04
347	0.699 ± 0.007	-0.306 ± 0.019	0.699 ± 0.007	0.393 ± 0.02
D_2 system				
273	1.370 ± 0.02	-0.265 ± 0.06	1.370 ± 0.02	1.105 ± 0.06
303	1.126 ± 0.08	-0.292 ± 0.09	1.126 ± 0.08	0.834 ± 0.10
347	0.890 ± 0.09	-0.340 ± 0.12	0.890 ± 0.09	0.550 ± 0.15

[a] The regression equation describing the composition dependence of the relative constants at each temperature is as follows: $K(X, T) = Xm(T) + b(T)$.

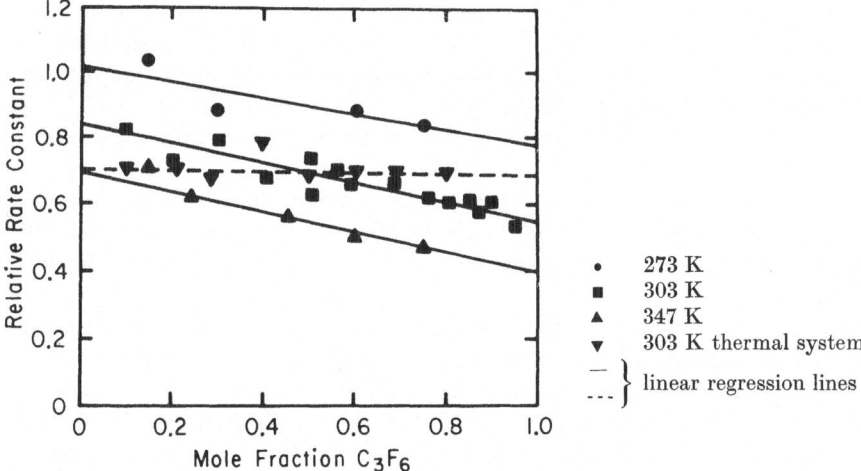

Fig. 4.2.1. Composition dependence of relative rate constants in H_2/C_3F_6 systems (taken from Ref. 4.2.5)

Ion-Molecule Reactions [4.2.6]

The basic roles of ion-molecule reactions have been realized in many interesting fields such as atmospheric-, discharge-, and radiation-chemistry. Recent developments in the chemistry of gaseous ions have been coupled with increasing availability of various new experimental techniques such as high pressure mass spectrometers, ion-cyclotron resonance mass spectrometers and tandem mass spectrometers, and have already provided a rich body of fundamental results. However, even with these sophisticated applications, mass spectrometric approach is still inadequate in many respects for solution chemistry.

The use of labeled gaseous ions from radioactive decay processes has proved to be helpful to perform the study of electrophilic reactions in gaseous phase. At least in a few typical cases, the technique will fill the gap between solution chemistry and mass spectrometry, since it not only allows us to study the reactions at high pressures but also yields information regarding the structure of the product molecules.

Ions from T-decay. In the decay of an isolated T atom, the excitation of the resulting $^3He^+$ ion is mainly due to momentum transfer from the emitted β-particle and anti-neutrino, and the "shaking" of the electron cloud due to the sudden increase in the nuclear charge. The theoretical evaluation [4.2.7] indicates that the decay leads predominantly to the formation of $^3He^+$ in its ground state (70%), with a significant fraction in excited states.

In the decay of T in chemical combinations, however, the survival of the resulting ^3He—X bond (X = C, N, O, F and others) [4.2.8] can be determined on the basis of the excitation energy and the potential energy surface of the bond, and its extent has been studied both theoretically and experimentally. The results are consistent with assumption that the daughter species are formed with very little internal excitation energy, and that the decay of tritiated hydrocarbons is always followed by the loss of neutral ^{3}He from the daughter ion: only about 20% of the primary organic ions undergoes further fragmentation, although over 90% of T_2 (or HT) decay events leads to the formation of stable ^{3}HeT$^+$ (or ^{3}HeH)$^+$ daughter ions.

The resulting organic ions can be used as a convenient source for the studies of cationic reactions both in the gas and liquid phases. Versatile results have been produced through experiments in which molecules labeled with more than two tritium atoms within the molecules have been used. In such cases, after decay of one tritium, the other serves as the tracer for the fate of the daughter species.

However, one may encounter difficulties both in synthesizing precursors with at least two tritium atoms in equivalent positions, and in minimizing the radiolysis due to β-particles emitted in the decay of tritium. With sophisticated techniques for synthesis and careful control of the reaction systems, these difficulties have been overcome and a wealth of information has been obtained regarding the reactions of various cations, ^{3}HeT$^+$ and R$^+$, resulting from the T-decay of suitable tritiated precursors.

^{3}HeT$^+$ has been used conveniently as a powerful proton donor for various organic compounds, resulting in the formation of the corresponding carbonium ions. For example, in the reaction with c-C_4H_8 [4.2.9], the observed C_4H_7T will result from the following reaction schemes:

$$^3HeT^+ + c\text{-}C_4H_8 \rightarrow [C_4H_8T]^{*+} + {}^3He, \tag{4.2.6}$$

$$[C_4H_8T]^{*+} + M \rightarrow (C_4H_8T)^+ + M^*, \tag{4.2.7}$$

$$(C_4H_8T)^+ + c\text{-}C_4H_8 \rightarrow C_4H_9{}^+ + C_4H_7T \begin{cases} \text{cyclobutane-T} \\ \text{t-2-butene-T.} \end{cases} \tag{4.2.8}$$

Labeled alkyl and cycloalkyl ions have been obtained from the decay of the corresponding multi-tritiated hydrocarbons, such as CT_3^+ ion from CT_4, $C_2H_4T^+$ from 1,2-$C_2H_4T_2$, c-$C_4H_6T^+$ from c-$C_4H_6T_2$, and allowed to react with various reactants, including hydrocarbons, halogen derivatives and alcohols.

In the reaction of CT_3^+ with benzene, toluene and their mixtures both in the gas (from 36 to 350 Torr) and liquid phase [4.2.10], the major reaction channels are methylation, substitution of hydrogen atom by tritium and substitution of methyl group by CT_3, occurring to different extent in gaseous and liquid systems. Tritiated toluene and xylene are formed in gaseous toluene at 36 Torr in the ratio of 1.2:1.0 with an overall absolute yields of about 5%. In the reactions with halobenzenes [4.2.11], the formation of halotoluenes, methyl halide, and toluene are the three reaction channels. The positional selectivity in the substitution for H atom in the liquid phase is almost the same for three halogens, C_6H_5X (X = F, Cl, Br); 40% for otho, 25% for meta, and 36% for para-positions.

The potential mechanistic interest of these results stems from the fact that cations are unsolvated and absolutely free. Thus the studies provide definite data regarding the role of the carbonium ions in aromatic alkylation.

Halogen Ions. The role of ionic species in the chemical processes following nuclear transformation of halogen atom has long been recognized, and the experimental techniques for the study of ionic reactions of gaseous halogen ions have also been described [4.2.10]. The idea is based on the fact that the ^{80m}Br(I.T.)^{80}Br or ^{125}Xe(E.C.)^{125}I reaction provides ^{80}Br or ^{125}I in various states of charge, electronic and translational energies, via Auger processes and vacancy cascades. When the decay occurs in sufficiently large excess of a rare gas, whose ionization potential lies between the first and the second ionization potential of Br or I, the charge of ^{80}Br or ^{125}I is finally reduced to $+1$ state in thermal energy range. The radioactivity of the daughter ion, ^{80}Br or ^{125}I, has an advantage over the β-decay of T or ^{14}C, such that the daughter acts itself as a radio-tracer and the use of the double-labeling technique is no longer necessary. As a typical example, gas phase aromatic bromination and iodination have been studied in C_6H_5X (X = F, Cl, Br, and CH_3) by this decay technique [4.2.12, 13]. Argon or Xe was used as diluent.

$$CH_3^{80m}Br \xrightarrow{\text{(I.T.)}} [^{80}Br]^{n+}_{ex} \xrightarrow{\text{collisions}} {}^{80}Br^+ \qquad (4.2.9)$$

$$^{80}Br^+ + C_6H_5X \rightarrow [C_6H_5X^{80}Br]^+_{ex} \qquad (4.2.10)$$

$$[C_6H_5X^{80}Br]^*_{ex} \longrightarrow \text{fragments} \qquad (4.2.11\,a)$$
$$\xrightarrow{M} [C_6H_5X^{80}Br]^+_{ex} + M^* \qquad (4.2.11\,b)$$

$$[C_6H_5X^{80}Br]^+ + C_6H_5X \rightarrow C_6H_4X^{80}Br + C_6H_6X^+. \qquad (4.2.12)$$

The results indicate that the positional selectivity is very low by the standard of solution chemistry. Furthermore, substitution by ^{80}Br$^+$ and ^{125}I$^+$ takes place for both hydrogen and substituent, X.

The ease for replacement of X relative to that of hydrogen increases from fluorobenzene to bromobenzene, with a concomitant increase in relative importance of ortho-substitution, as seen in Table 4.2.3.

Table 4.2.3. ^{80}Br-for-H and ^{80}Br-for-X substitution (taken from Ref. 4.2.13; Copyright 1974, American Chemical Society)

Product	Substrate				
	C_6H_6	C_6H_5F	C_6H_5Cl	C_6H_5Br	$C_6H_5CH_3$
o-C_6H_4XBr		1.02	0.84	0.36	1.90
m-C_6H_4XBr		1.10	0.79	0.27	1.45
p-C_6H_4XBr		1.11	0.85	0.28	1.19
C_6H_5Br	3.00	0.43	2.16	18.06	1.25
Br-for-X : ΣBr-for-H		0.13	0.87	19.80	0.27
$1/_2$ ortho : para		0.46	0.49	0.64	0.80
$1/_2$ meta : para		0.50	0.46	0.48	0.61

a In per cent of total ^{80}Br daughter ions formed; average of at least three individual runs, whose standard deviation is about 10%.

b The gaseous system was kept at 760 Torr at 70°; composition: 95 mol % of Ar, 2.5 mol % of C_6H_5X, and 2.5 mol % of ^{80m}Br-labeled CH_3Br as ^{80}Br$^+$ source.

Ions from ^{14}C*-decay.* The theoretical treatment of the consequence of the ^{14}C-decay in molecules predicts that the retention values in different molecules are close to 70%, nearly irrespective of the molecular structure [4.2.14]. This is in fair agreement with the experimental observation that more than 50% of the ion from the decay

$$R_1{}^{14}CH_2R_2 \xrightarrow{\beta\text{-decay}} (R_1R_2NH_2)^+ \tag{4.2.13}$$

will survive further fragmentation.

Although no report has been made specifically on the further ion-molecule reactions of these primary ions, the β-decay of ^{14}C-labeled precursors will have a potential value as a useful source of cations containing nitrogen atom.

References

[4.2.1] Tang, Y. N.: *Isotopes in Organic Chemistry* Vol. 4, E. Buncel, C. C. Lee (eds.), Elsevier 1978, p. 85
[4.2.2] Fee, D. C., Markowitz, S.: J. Phys. Chem. *78*, 347 (1974)
[4.2.3] Rowland, F. S.: private communication
[4.2.4] Williams, R. L., Rowland, F. S.: J. Phys. Chem. *77*, 301 (1973)
[4.2.5] Grant, E. R., Root, J. W.: J. Chem. Phys. *64*, 417 (1976)
[4.2.6] Cacace, F.: Hot Atom Chemistry Status Report, IAEA, Vienna, (1975), p. 229
[4.2.7] Migral, A.: J. Phys. (USSR) *4*, 449 (1941)
[4.2.8] Ikuta, S., Yoshihara, K., Shiokawa, T.: J. Nucl. Sci. Technol. *14*, 720 (1977)
[4.2.9] Cacace, F., Guarino, A., Passagno, E.: J. Am. Chem. Soc. *91*, 3131 (1969)
[4.2.10] Cacace, F.: Abstract of the 9th Int. Symp. Hot Atom Chem., Blacksburg, Virginia, U.S.A. (1977), p. 81
[4.2.11] Giacomello, P.: ibid., p. 82
[4.2.12] Cacace, F., Stocklin, G.: J. Am. Chem. Soc. *94*, 2518 (1972)
[4.2.13] Kunst, E. J., Halpern, A., Stöcklin, G.: ibid. *96*, 3733 (1974)
[4.2.14] Raadschelders-Buijze, C., Roos, L. L., Ros, P.: Chem. Phys. *1*, 468 (1973)

4.3 Applications in Biochemistry and Nuclear Medicine

There are two important aspects regarding hot atom chemistry in the field of biochemistry. One is production of labeled compounds, especially radiopharmaceuticals. The other is concerned with chemical and biological effects of radionuclides in human body. Since the successful application of recoil technique in preparing nearly carrier-free ^{82}Br-methyl bromide [4.3.1], various labeled compounds have been produced by similar procedures. Although the majority of the labeling works have been done with a view to understanding the mechanisms involved [4.3.2—5], it is worth while to consider the present situation in production of radiopharmaceuticals in relation to hot atom chemistry.

Radiopharmaceuticals

The term "radiopharmaceuticals" is applied to a wide variety of substances used in biological- and medical researches and in clinical applications. However, they are generally defined as radioactive materials which are administered to man for therapeutic or diagnostic purposes and which therefore undergo metabolic changes in the organism [4.3.6].

At present, main demands for radiopharmaceuticals arise in their uses as radioactive tracers in diagnosis. Useful information is available on the morphological diagnosis of certain organs and systems, and uptake of materials in various organs.

Appropriate radiopharmaceuticals have to be chosen so that they will be concentrated in the organ of interest according to the physiological characteristics. The rapid excretion after the diagnosis is also desirable. Radiopharmaceuticals injected into a human body either decay according to their own half-lives, or are excreted as the consequence of biochemical metabolism. The effective half-life in the body can be taken as a measure of the internal dosage, and nuclides with shorter half-lives are more preferable on the condition that they should survive long enough to work effectively for the original purposes. Another important advantage of a shorter lived nuclide is that specific activity can be made high enough to act as "true" tracer. Otherwise, amounts of non-active compounds required become large, and the concomitant toxicity or physiological response may perturb the original purpose for which the tracer is used. Since it is essential to maintain the administered amount below a limited value, recent developments in radiopharmaceuticals have been specially aimed at the production of useful radionuclides with shorter half-lives [4.3.7—9, 21]. This is closely related with the availability of various charged particles from accelerators. On introducing charged particles onto the nucleus, the radioisotopes produced are situated on the neutron-deficient side of the nuclear stability curve and tend to decay by either electron capture or positron emission which are more favored in nuclear medicine. In positron emission, the positron annihilation provides advantages over single photon imaging. i.e., increased sensitivity and availability of a three dimensional imaging.

Labeling Methods

Two common procedures widely adopted for preparing radiopharmaceuticals are the chemical synthesis and bio-synthesis. In several cases, however, non-synthetic techniques such as recoil method, radiolytic method, labeling by exposure to atomic species, accelerated ion method, and excitation labeling method can also be used to produce the product directly for immediate use or for the use after isolation and purification [4.3.7, 24]. In the excitation labeling, the precursor of the labeling nuclide should have a convenient half-life and charge state and excitation of the labeling nuclide provides its "labeling power". As an example of this method, the decay of ^{123}Xe in Cl_2 gas has resulted in 100% yield of $Cl^{123}I$. When the decay occurs in indocyanine green crystals, non-specific labeled product has resulted in 20% yield, in potential use as a liver function agent. Recently Juses et al. [4.3.35] have studied the labeling with ^{80}Br or ^{82}Br by excitation labeling. The recoil method provides the simplest means to produce labeled radiopharmaceuticals, and can often be used particularly when no other methods provide the desired compounds. Serious disadvantages, more or less common to the other non-synthetic methods, include (i) random labeling of the nuclide in the molecule and (ii) relatively small yields due to the occurrence of other competing reactions.

In labeling with elements other than carbon and tritium, random labeling becomes less important since it gives chemically different products, of which mutual separation is usually attained. The ^{18}F-fluoroacetic acid with 9—13% yields has been obtained with the ^{19}F$(\gamma, n)^{18}$F reaction in fluorocarbon-11($CFCl_3$)-glacial acetic acid systems [4.3.10].

However, in labeling carbon or tritium, random labeling limits the applicability of the recoil method to such cases only when one wishes to trace a molecule as a whole but not any particular atom in the molecule. When a complex molecule is chosen as a reactant in reactions with recoil carbon or tritium, there exist various reacting sites in the molecule and random labeling occurs. In the reaction of recoil carbon with organic compounds, two interesting products are "re-entry" products and "synthesis" products [4.3.4]. However, the labeling in these products is neither uniformly random nor specific.

When a molecule with a single reactive site toward the recoil atoms is chosen as a reactant, the product is labeled only at single position. For example, in the reactions of recoil ^{11}C with B_2O_3, only ^{11}CO and ^{11}CO$_2$ are formed and easily separated from each other; they can be used as intermediates in the synthesis of desired final compounds. This leads to a novel approach to recoil labeling by preparing an intermediate for the synthesis by recoil method. At present, relatively few precursors are available for each nuclide (Table 4.3.1), in which there is much hot atom chemistry. Direct preparation of intermediates of posirion emitters will be briefly described in the following sections.

Carbon-11. ^{11}C-carbon monoxide, ^{11}C-carbon dioxide and ^{11}C-cyanide as H^{11}CN or Na^{11}CN are the three ^{11}C-precursors most commonly used in the synthesis of ^{11}C-compounds. Among the nuclear reactions producing ^{11}C, the three listed below are in routine use for production purposes:

$$^{10}B(d, n)^{11}C, \quad ^{14}N(p, \alpha)^{11}C, \quad \text{and} \quad ^{11}B(p, n)^{11}C.$$

^{11}C-carbon monoxide and carbon dioxide can be easily prepared in multicurie

Table 4.3.1. Precursors for the synthesis of compounds labeled with positron emitters (taken from Ref. 4.3.8)

Carbon-11	CO CO$_2$ CN$^-$ CH$_2$O CH$_3$I CH$_3$OH
	C$_2$H$_2$ C$_2$H$_4$ C$_2$H$_5$OH CH$_3$—COOH $\overset{\displaystyle O}{\overset{\displaystyle \|}{NH_2-C-NH_2}}$
Nitrogen-13	NO NO$_2$ N$_2$O N$_2$ HCN NH$_3$
Oxygen-15	O$_2$ H$_2$O CO CO$_2$
Fluorine-18	F$_2$ F$^-$
Salts of fluorine	(C$_2$H$_5$)$_4$N$^+$F$^-$ or crown ether salts NOF ClF CF$_3$OF (C$_2$H$_5$)$_2$NSF$_3$ SbF$_3$ LiBF$_4$ FClO$_3$ BiF$_3$ etc.

amounts by bombarding B_2O_3 target with a low energy deuteron beam. The relative composition of $^{11}CO/^{11}CO_2$ is mainly determined by the target sweep gas composition [4.3.26]. The use of hydrogen inhibits the formation of $^{11}CO_2$. ^{11}C-carbon monoxide is presumably the primary product, which is then oxidized to $^{11}CO_2$ by oxygen atoms produced from the boric oxide by the radiation-induced reaction.

Lamb et al. [4.3.11] have described the preparation of ^{11}C-cyanide by two different methods. In the first method, 15 MeV proton irradiation of $LiNH_2$ results in the formation of $Li^{11}CN$ as the consequence of recoil reactions. In the other, $H^{11}CN$, produced directly from proton bombardment of a $99\%N_2-1\%H_2$ mixture was oxidized by $KMnO_4$ to ^{11}C-cyanide. The modification of the method improved the $H^{11}CN$ yield to above 95% of the ^{11}C produced in the target [4.3.7]. When the radiation dose in the system is greater than 0.5 eV/molecule, the $H^{11}CN$ is quantitatively converted to $^{11}CH_4$, with simultaneous formation of NH_3. When this gas mixture is passed over platinum wool at $1000\,^\circ C$, the $^{11}CH_4$ is quantitatively converted back to $H^{11}CN$,

$$NH_3 + {}^{11}CH_4 \xrightarrow[1\,000\,^\circ C]{Pt\ wool} H^{11}CN\,.$$

^{11}C-acetylene can be prepared either via conversion of ^{11}CO to $H^{11}C{\equiv}CH$ through the formation of $Ba^{11}CO_3$ or direct recoil labeling of the carbides [4.3.36]. An alternative method has been developed, which provides carrier-free ^{11}C-acetylene by using cyclopropane as target:

$$c\text{-}C_3H_6 \xrightarrow{(p,pn)} H^{11}C{=}CH\,.$$

The radiochemical yield of ^{11}C-acetylene was 50% [4.3.7]. ^{11}C-methyl iodide was prepared in yields up to 25% by the on-line bombardment of N_2-HI mixtures with protons. It is further reported that ^{11}C-methylamine with 45% radiochemical yield and ^{11}C-guanidine with 49% radiochemical yield can be obtained from the proton bombardment of crystalline NH_4Cl and NH_4I, respectively [4.3.7].

A novel method for the synthesis of another important intermediate, ^{11}C-phosgene, has been reported using high pressure nitrogen target [4.3.25]. The method involves an improved catalytic reduction method for preparation of ^{11}CO and a 4π-geometry low pressure mercury lamp. The radiochemical yield for the reaction, $^{11}CO + Cl_2 \xrightarrow{U.V.} {}^{11}COCl_2$, is nearly quantitative within five minutes.

Fluorine-18 [4.3.18]. The particular importance of fluorine-18 in radiopharmaceuticals is the fact that fluorinated organic compounds often results in similar or enhanced biological activity to the influorinated analog. Thus, whenever synthesis of the desired compound is difficult, ^{18}F readily offers an alternative as a tag provided the labeled position does not inhibit biological activity. By far the most common reactions in use are the $^{16}O(^3H, n)^{18}F$, and $^{16}O(^4He, pn)^{18}F$, where the target can be water or oxygen gas. The neon reaction, $^{20}Ne(d, \alpha)^{18}F$, proves to be one of the most versatile for fluorine-precursor production. Rambrecht et al. [4.3.12] have irradiated with deuterons various mixtures of Ne and an additive (Fe, H_2, Cl or CF_2O) in high purity nickel targets passivated with anhydrous film of nickel fluoride for preparing usable carrier-free anhydrous fluorinating agents, such as [18]FF, H[18]F, Cl[18]F and C[18]FF$_3$.

Pure CF_3OF can be obtained by fluorination of carbon monoxide, carbon dioxide, methanol, or carbonyl fluoride at high temperatures in copper vessels, or by catalytic processes. However, recent demands for the availability of [18]F—CF_3OF in radiopharmaceuticals request convenient and rapid preparation of this compound. Neirinckx et al. [4.3.13] have found CsF as a superior catalyst. They have bombarded with deuterons Ne gas in a NiF-coated target vessel: dried, powdered CsF was spread over the bottom of the vessel. The ^{18}F atoms produced were trapped in CsF through an exchange reaction. After evacuation, F_2 and CF_2O were added and equilibrated. A 35-min reaction at $100\,°C$ converted 97% of the CF_2O to CF_3OF. Roughly one-third of ^{18}F in [18]F—CF_3OF exists in the —OF group.

Nitrogen-13 [4.3.14]. The $^{12}C(d, n)^{13}N$ reaction is most commonly used in production of ^{13}N-labeled gases, and various targets graphite, activated charcoal, CO_2, etc.) are chosen. In the bombardment of CO_2 [4.3.15], ^{13}N atoms formed may react with traces of N_2 impurity, and mostly result in ^{13}NN.

^{13}N-ammonia has come to be used not only in myocardial imaging, but also as a synthetic precursor mainly in production of ^{13}N-aminoacids. Bombardment of an adequate carbide, such as Al_4C_3, can produce large amounts of $^{13}NH_3$ with some methylamine [4.3.16]. An alternative procedure for production of $^{13}NH_3$ involves the deutron bombardment of methane: methane in an open circuit flowed through a Pyrex-glass lined tube of the target chamber into water or isotonic saline to dissolve ammonia [4.3.17]. The $^{16}O(p, α)^{13}N$ reaction is also used, and Vaalburg et al. obtained with the proton beam of about 19 MeV a high yield of $^{13}NH_3$ as shown in Table 4.3.2 [4.3.8].

Table 4.3.2. Comparison of $^{13}NH_3$ yields from various production methods (taken from Ref. 4.3.18)

Nuclear reaction	Target material	Particle energy (MeV)	Radiochemical purity (%)	Yield mCi/μA/20 min
$^{12}C(d, n)^{13}N$	CH_4	8	80—95	4.4
$^{16}O(p, α)^{13}N$	O_2	< 15	99.6	5.6
	H_2O	< 15	100	15
	H_2O	19	99.9	36

In the proton bombardment of high pressure O_2, ^{13}N atoms are incorporated into the primary products, $^{13}N_2$, $^{13}N_2O$ and $^{13}NO_2$ in the ratio 8:1:3.5 [4.3.24]. ^{13}N-nitrous oxide, which has been used to assess total cerebral blood flow, was synthesized [4.3.20] by a pair of processes based on pyrolysis of NH_4NO_3 in sulfuric acid. Both methods start with $^{13}NO_3^-$ production via the proton irradiatoin of water, and NH_4NO_3 is added later.

Oxygen-15. Oxyhemoglobin and carboxyhemoglobin are labeled by dissolving $O^{15}O$ and $C^{15}O$ respectively in blood. The short half-life (2.04 min) proves to be an advantage for in vivo investigation regarding cerebral metabolism, permitting rapid sequential injections. ^{15}O-labeled molecular oxygen has been prepared by deutron irradiation of continuously flowing mixtures of N_2 and O_2. Most of ^{15}O (97.5%) was tagged in O_2 with the gas mixture free from CO_2 [4.3.22]. This labeled gas can then be converted to $C^{15}O$ and $CO^{15}O$ by passage over activated charcoal at 900 and 400 °C. Another route to $CO^{15}O$ is via the irradiation of N_2-CO_2 gas mixtures. By bubbling $CO^{15}O$ gas through water, $H_2^{15}O$ is formed through the equilibrium, $H_2O + CO^{15}O = H_2^{15}O + CO_2$. The direct recoil synthesis has recently been attempted by deutron bombardment of H_2-N_2 mixtures [4.3.23]. $H_2^{15}O$ with high specific activity has been obtained.

Biological Effects

Nuclear transformations of radionuclides incorporated into biological systems may alter effectively the structures and functions of vitally important organic compounds.

The multitude of organic compounds in any living structure include both molecular species being rapidly turned over within cells and those being only slowly or not turned over. The latter are practically irreplaceable for the cellular system, and the damage to these has functional implication. The genetic material, deoxyribonucleic acid (DNA), is classified in this group, and it is considered in radiobiology as the critical molecule in the cell.

The molecular alteration of DNA has been widely examined as the consequence of radiation. Single- and double-strand breaks, base alterations and interstrand cross linking occur. Single-strand breaks and base changes are usually efficiently repaired in the living cell, whereas double-strand breaks are only inefficiently repaired.

A difference exists between the radiation-induced effects and those associated with transmutation of radionuclides. In the former, the effects are randomly distributed in the cellular system, while in the latter, transmutation events occur at a specific site where the radionuclide is bound. Primary factors contributing to the transmutation effects are as follows:

(i) The chemical change of the nuclide at the time of transmutation.

(ii) The residual electronic excitation of the newly formed nuclide.

(iii) The recoil of the transmuted nuclide.

In addition to these, radiations emitted in the transmutation events add further effects onto the system. Although it is not easy to distinguish between the two contribution to overall effects, the effects from the absorbed radiation usually outweigh the transmutation effects, particularly in transmutation of β-emitting nuclides.

Because of the specific role of DNA in the living systems, emphasis has been placed on the effects of radionuclides which may be incorporated into DNA under various circumstances. Thus, both β-emitting nuclides (e.g., T, $^{32,33}P$, ^{14}C) and those decaying by electron capture (e.g., ^{125}I) are included in the examinations.

Table 4.3.3. Percentage of single-base substitutions produced by mutagens listed (taken from Ref. 4.3.30)

Base substitution	Mutagen					
	Thymidine-methyl-³H	Uracil-6-³H	Histidine-³H	Ionizing Radiation	Uracil-5-³H	Thymidine-2-¹⁴C
GC—AT	24	24	18	17	96	11
GC—CG	2	4	12	6	2	4
GC—TA	45	43	41	45	1	13
AT—GC						
AT—CG	9	15	9	6	1	50
AT—TA						

On the basis of the published data, the emitted β-particles are mainly responsible for the observed biological effects in the β-emitting nuclides [4.3.27—29]. The effects are usually localized and their extents are very sensitive to the labeling position of the nuclide. For example, in transmutation of T the base change [4.3.30] induced by T-decay in 5-position of the cytosine in DNA is significantly higher than that found in T-decay in the 6-position, in the methyl group of thymidine, or in histidine, as seen in Table 4.3.3. In the Table it is further noticed that the data from the latter three positions are very similar to those from ⁶⁰Co γ-rays.

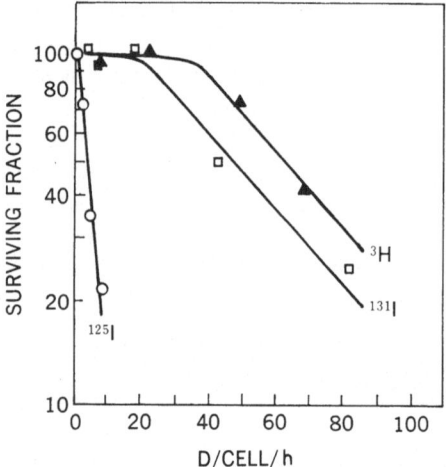

Fig. 4.3.1. Dose-survival curves for ¹²⁵I, ¹³¹I, and ³H (taken from Ref. 4.3.31)

Contrary to the importance of radiation emitted in the transmutation of β-emitting nuclides, the transmutation effect of ¹²⁵I appears to be much more pronounced and outweigh the effect from the emitted radiations. When nuclides such as ¹²⁵I decay by electron capture, inner shell vacancies are rearranged due to Auger effect, and thus several electrons are emitted per decay. Because of the short range of Auger electrons, the particle energy is absorbed mostly in the close vicinity of the decaying nuclide. Assuming a random distribution of

DNA molecules labeled with ^{125}I in cell nuclei, the amount of energy absorbed in ^{125}I-decay within a spherical volume of 8 μ in diameter has been evaluated, as compared with that in T-decay. The calculation indicates that the energy absorption in ^{125}I-decay amounts to 3.5 times as much as that in T-decay. In spite of such value expected from the absorbed dose, experiments with mouse leukaemia cells labeled with ^{125}I-iododeoxyuridine, ^{131}I-iododeoxyuridine, or T-thymidine revealed that ^{125}I-decay was much more effective on cell killing than [131]I- or T-decay, as shown in Fig. 4.3.1 [4.3.31]. The enormous biological effectiveness of ^{125}I decay observed in DNA can be related with the effective production of double-strand breaks by the decay. Schmidt et al. [4.3.32] have analyzed the rate of double-strand breaks in the T-phage labeled with ^{125}I-iododeoxyuridine and found that every second ^{125}I-decay is effective in producing the double-strand breaks. Hence, a full explanation of the effects of ^{125}I-decay should involve the mechanisms other than those initiated by ionization and excitation from electron irradiation, i.e., the fragmentation of molecules due to Coulombic repulsion. A clear-cut experiment has been conducted to demonstrate local destruction of molecules following the Auger process. When a zinc-containing enzyme, carbonic anhydrase, was irradiated with monoenergetic photons of discrete energies ranging from about 8 to 10.7 keV, a sudden rise in enzyme inactivation and a concomitant increase in zinc release occur at the energy of 10.01 keV due to Auger effect initiated by photoelectric effect in the K-shell of zinc, as seen in Fig. 4.3.2 [4.3.33].

In the irradiation of bromodeoxyuridine incorporated in DNA, a sudden increase in radical yields has been similarly noticed at the energy level corresponding to the K-absorption edge for Br atom. Deutmann and Stöcklin [4.3.34]

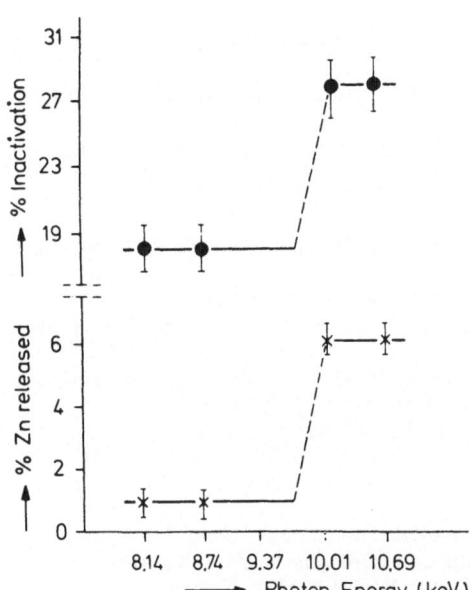

Fig. 4.3.2. Inactivation and zinc release in carbonic anhydrase (Reprinted with permission from Ref. 4.3.33. Copyright (1976) American Chemical Society)

have further examined the role of transmutation effects of ^{125}I in an iodinated DNA precursor, iodouracil, by using a double-labeling technique. The ^{125}I in [2-^{14}C, 5-^{125}I]-iodouracil was allowed to decay for several months in aqueous ethanol. An essential difference exists between decay effects and external γ-radiolysis. Gamma-radiolysis yields uracil as the major product. In the decay of ^{125}I, however, a complete disruption of the molecule occurs as the major reaction route and products with intact uracil-ring structure were not observed practically: $^{14}CO_2$ and ^{14}CO were two specific products.

The Auger process following radioactive decay or inner shell ionization is relevant to the fields: (i) selective microsurgery in biological macromolecules, (ii) radiotoxicity estimate of incorporated radionuclides, and possibly (iii) radiation therapy. Further elucidation of the mechanisms involved will be necessary in the future.

References

[4.3.1] Glueckauf, E., Jacobi, R. B., Kitt, G. P.: J. Chem. Soc. 330 (1949)

[4.3.2] Urch, D. S.: MTP Int. Rev., Sci., Inorg. Chem. Ser. Two 8, 849 (1975)

[4.3.3] Tang, Y. N.: Isotopes in Organic Chemistry, Vol. 4, E. Buncel, C. C. Lee (eds.). Elsevier 1978, p. 85

[4.3.4] Wolf, A. P.: Ann. Rev. Nucl. Sci. 10, 29, Annual Review Inc., Palo Alto (1960)

[4.3.5] Rupp, A. E.: Radiopharmaceuticals and Labeled Compounds, Vol. 1, IAEA, Vienna, (1973), p. 223

[4.3.6] Cohen, Y.: Analytical Control of Radiopharmaceuticals, Proc. of a Panel, Vienna, (1969), p. 1

[4.3.7] Wolf, A. P., Christman, D. R., Fowler, J. S., Lambrecht, R. M.: Radiopharmaceuticals and Labeleds Compounds, Vol. 1, IAEA, Vienna, (1973), p. 345

[4.3.8] Wolf, A. P., Fowler, J. S.: BNL-25850 (1978)

[4.3.9] Quinn, J. L. E.: The Yearbook of Nuclear Medicine, Vol. 1. Chicago: Yearbook Medical Pub. 1978

[4.3.10] Donnerhack, A., Sattler, E. L.: Presented at 10th Int. Symp. Hot Atom Chem., Loughborough, 1979

[4.3.11] Lamb, J. F., James, R. W., Winchell, H. S.: Int. J. Appl. Radiat. Isotopes 21, 475 (1970)

[4.3.12] Lambrecht, R. M., Neirinckx, R., Wolf, A. P.: Int. J. Appl. Radiat. Isotopes 29, 175 (1978)

[4.3.13] Neirinckx, R. D., Lambrecht, R. M., Wolf A. P.: ibid. 29, 323 (1978)

[4.3.14] Straatman, M. G.: ibid. 28, 13 (1977)

[4.3.15] Welch, M. J.: Prog. Nucl. Med., Vol. 3. B. L. Holman (ed.). Basel, Karger, and Baltimore: University Park Press, 1973

[4.3.16] Welch, M. L., Lifton, J. F.: J. Am. Chem. Soc. 93, 3385 (1971)

[4.3.17] Tilbury, R. S., Dahl, J. R., Manahan, W. G., Laughlin, J. S.: Radiochem. Radioanal. Lett. 8, 317 (1971)

[4.3.18] Vaalburg, W., Kamphuis, J. A. A., Molen, H. D. B., Reiffers, S. Rijskamp, A., Woldring, M. G.: Int. J. Appl. Radiat. Isotopes 26, 316 (1975)

[4.3.19] Parks, N. J., Peek, N. F., Goldstein, E.: ibid. 26, 683 (1975)

[4.3.20] Nickles, R. J., Gatley, S. J., Hichwa. R. D., Simpkin, D. J.: ibid. 29, 225 (1978)

[4.3.21] Silvester, D. J.: Radiopharmaceuticals and Labeled Compounds, Vol. 1, IAEA, Vienna, (1973), p. 197

[4.3.22] Welch, M. J., Ter-Pogossian, M. M.: Radiat. Res. 36, 580 (1968)

[4.3.23] Ruiz, H., Wolf, A. P.: J. Label. Comp. and Radiopharm. 15, 185 (1978)

[4.3.24] Wolf, A. P.: Hot Atom Chemistry Status Report, IAEA, Vienna, (1975), p. 271

[4.3.25] Christman, D. R., Finn, R. D., Wolf, A. P.: Presented at 10th Int. Symp. Hot Atom Chem., Loughborough, 1979
[4.3.26] Clark, J. C., Buckingham, P. D.: Radiochem. Radioanal. Lett 6, 281 (1971)
[4.3.27] Duplan, J. F., Chapiro, A. (eds.): Advances in Radiation Research. New York: Gordon and Breach Science Pub. 1973
[4.3.28] Feinendegen, L. E.: Hot Atom Chemistry Status Report, IAEA, Vienna, (1975), p. 285
[4.3.29] Biological Effects of Transmutation and Decay of Incorporated Radionuclides, Proc. of a Panel, Vienna, (1967)
[4.3.30] Person, S., Phillips, S. L., Anderson, F. A., Newton, H. P.: Int. J. Rad. Biol. 21, 159 (1972)
[4.3.31] Hofer, K. G., Hughes, W. L.: Rad. Res. 47, 94 (1971)
[4.3.32] Schmidt, A., Hotz, G.: Int. J. Rad. Biol. 24, 307 (1973)
[4.3.33] Diehen, D., Halpern, A., Stöcklin, G.: J. Am. Chem. Soc. 98, 1077 (1976)
[4.3.34] Stöcklin, G.: Radiation Research, Proc. of the 6th Int. congress of Rad. Res. (1979), Tokyo, p. 382
[4.3.35] Jesus, O. D., Mustaklem, J., Ache, H. J.: Presented at 10th Int. Symp. Hot Atom Chem., Loughborough, 1979
[4.3.36] Myers, W. G.: Nucl. Med. 13, 699 (1972)

4.4 Hot-Atom Chemistry in Energy-Related Research

Recent progress in the application of nuclear energy to human life requires further contribution of chemistry in which hot-atom chemistry will largely participate.

Chemical behaviors of fission-produced radionuclides in, and their release behaviors from nuclear fuel pins are particularly important performance-limiting factors in connection with the deformation of the integrity of the fuel pins and from the view-point of the release of radionuclides into the atmosphere. Furthermore, in fusion reactions, the recovery of tritium from blanket materials and the damage of constructing materials due to sputtering effects of plasma ions should be the objects for the hot-atom chemistry, which are essential for successful developments of such reactors.

Fission Reactors

Since one of the limiting factors for the use of UO_2 as nuclear fuel is the behavior of gaseous fission products, the diffusion characteristics of volatile fission products have been investigated extensively.

Release of Rare Gases. The experiments are classified into three major categories: post-irradiation annealing experiments, in-pile experiments, and others including ion-bombardment and doping in the fabrication stage of samples [4.4.1].

It is known that fission product gases are released from the fuel through a combination of mechanisms: direct recoil release, knock-on release, and thermal diffusion. In the first mechanism, nuclides leave the UO_2 surface as the consequence of the recoil energy, and the release rate is proportional to the amount of nuclides formed within the depth of recoil range below the surface. In the knock-on release, emission of nuclides with excess kinetic energy is accompanied by a mass of surface molecules (about 2000 UO_2 molecules on average). Fission gas contained within the knocked-out volume will also be released together with the UO_2 molecules. Since the equilibrium concentration of nuclides in the knocked-out zone is proportional to the fission rate, the overall release rate is related with the square of the neutron flux. The dominant mechanism associated with normal operating temperatures, however, is the diffusion of gas atoms to grain boundaries followed by the growth of channels leading to open surfaces. Booth [4.4.2] has introduced a mathematical model for fission-product gas release and obtained solutions based on Fick's law:

$$\frac{\partial C}{\partial t} = D\nabla^2 C \qquad (4.4.1)$$

where C is the fission gas atom concentration, and D the classical diffusion coefficient. The approximate solutions for a spherical geometry give the fractional release (F) of the fission product gas as

$$F = \frac{6}{a}\sqrt{\frac{Dt}{\pi}} - \frac{3Dt}{a^2}, \qquad \left[\frac{\pi^2 Dt}{a^2} \leq 1\right] \qquad (4.4.2)$$

where a is the radius of an equivalent sphere. Later, it has become clear that the fission gas atoms may not migrate solely by classical diffusion. The "radiation enhancement" is one of the phenomena which cannot be explained: the values obtained from irradiation at low temperatures were higher than those predicted by extrapolating the results from the post-irradiation experiments. Such enhancement will result in either a temperature-dependent activation energy, or an activation energy lower than those obtained in the out-of-pile experiments. Regardless of these limitations, Booth's model has proved to be very useful and can still be employed if D is taken to be an effective value of diffusion, including the effects of capture and escape of fission gas products at various internal traps. The observed values for the diffusion coefficient have been widely scattered, and the variations in physical and chemical properties of the fuel and in the experimental conditions have been examined to account for such spread [4.4.1].

a) *Effects of Burn-up.* MacEwan and Stevens [4.4.3] have carried out post-irradiation annealing on carefully prepared single crystals of UO_2 and sintered UO_2. The apparent diffusion coefficient (D') was decreased rapidly in the range 4 to 20 MWD/te(U). Carroll and Sisman [4.4.5] have observed a similar trend in in-pile experiments even at $700\,^\circ C$. At these temperature, two mechanisms have been proposed, i.e., knock-out, and direct fission recil. The knock-out release is proportional not only to the square of the burn-up rate but also to the total macroscopic surface area. The observed burn-up effects have been attributed to the decrease in the total surface area. However, MacEvan and Stevens [4.4.3] have provided an alternative explanation that accumulated trapping sites may prevent gases from reaching the surface layer for subsequent release by the knock-out mechanism.

At present, the real reason for the decrease in D' is not clear yet. However, it is obvious that the reduction mechanisms reach saturation and D' assumes a smaller constant value after certain burn-up, probably in the region 70 to 400 MWD/te(U). A dramatic increase [4.4.6] in the release rate at much higher burn-up (18,000 MWD/te(U)) is explained in terms of internal cracking of the fuel due to accumulation of fission products: the cracking will enhance the release of stored fission gas and increase the surface area available for the gas release.

b) *Effects of Rating.* The results shown in Fig. 4.4.1 are summarized by Carroll et al. [4.4.5]. The release rate reveals a very characteristic dependence on rating. The deviation of the curve from classical diffusion has been explained in terms of the defect trap model, in which the increase in the rate of production of irradiation-

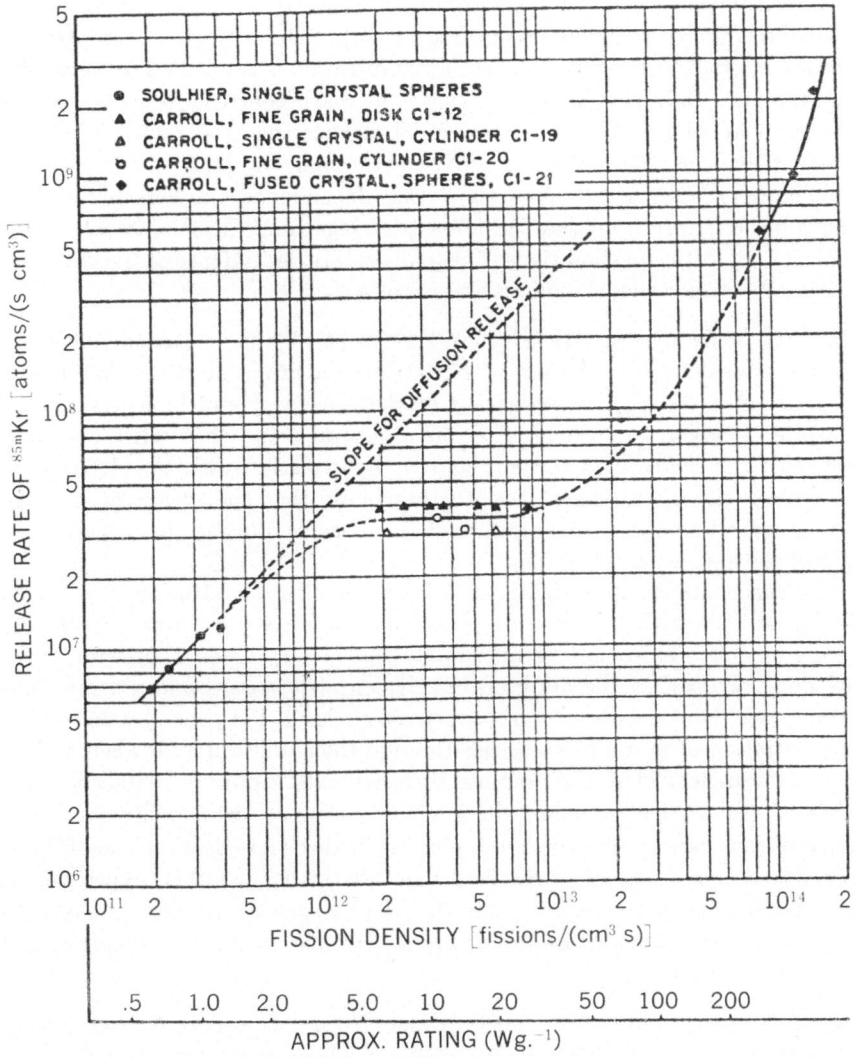

Fig. 4.4.1. Variation of release rate with rating (taken from Ref. 4.4.6)

induced trapping sites is assumed as sufficient to trap all extra fission gas atoms and bring about a decrease in the fractional release per unit time. However, as has been pointed out by Lawrence [4.4.1], the results in the figure should be corrected for the difference in the burn-up. In Souhlier's results, the maximum burn-up achieved was about 6 MWD/te(U), in contrast to 250 to 1000 MWD/te(U) in the experiments by Carroll et al. The correction for the different burn-up reduces the Soulhier's results by about an order of magnitude.

As for the "radiation enhancement of diffusion", Speight gives an expression for D as a function of rating [4.4.1]:

$$D = (1/2)\, D_T + (1/2)\, [D_T^2 + AM \cdot \exp(E/RT)]^{1/2}. \tag{4.4.3}$$

where D_T is the "thermal" out-of-pile diffusion coefficient, M the fuel rating, R the universal gas constant, T the absolute temperature, and A and E constants. They have further argued that fission fragments produce an extra concentration of vacancy-interstitial pairs and the gas atom diffusion is controlled by vacancies in the uranium sub-lattice. However, this is in sharp contrast to the argument by Carroll et al. [4.4.5], that the production of vacancies and point defects by fission fragments should reduce the release through trapping the gas atoms. Hence, the detailed mechanism of the rating dependence still remains unclear and certainly requires more detailed knowledge regarding the microscopic structure of UO_2 and diffusion conditions. The temperature dependence of the variation of release rate with rating should also be examined.

c) *Effects of Stoichiometry.* Miekeley and Felix [4.4.7] have carried out post-irradiation annealing experiments with UO_2 over a range of stoichiometry for the purpose of a detailed investigation of the effects of stoichiometry on the release rate of fission gas atoms. There is a general trend that diffusion is enhanced with the increase in oxygen content but coupled with the decrease in the activation energy, Q. They have further found that the sweep gas used in annealing can alter the stoichiometry of the sample during the experiments and affect the experimental results.

Most experiments for investigation of the mechanism of diffusion were based on doping of UO_2 with trivalent or pentavalent additives to create artificially an excess of anion or cation vacancies. Matzke [4.4.8] doped UO_2 with Y_2O_3, La_2O_3, ZrO_2, or Nb_2O_5. The addition of trivalent lanthanum and yttrium ions should increase the concentration of oxygen vacancies in UO_2, thereby decreasing the concentration of uranium vacancies through the equilibrium between cation- and anion-vacancies. Thus the uranium diffusion rate should be reduced, while addition of pentavalent niobium ions should enhance cation diffusion. The results, however, have indicated that doping has little or no effect on the diffusion of rare gas atoms. These observations together with the other experimental results [4.4.9] appear to demonstrate that fission gas atoms are unlikely to diffuse interstitially, but diffuse on a vacancy complex, e.g., a two-oxygen and one-uranium vacancy cluster.

Release of Tritium. In the UO_2 nuclear fuel, tritium is generated by ternary fission with thermal and fast fission yields of about 0.9×10^{-4} and 2×10^{-4} atoms/^{235}U fission respectively. The tritium formed diffuses through the matrix and is released into a plenum during the reactor operation. Penetration of tritium through cladding depends strongly on the kind of cladding material; with stainless steel more than 90% of tritium penetrates the cladding into coolant, while the major fraction of tritium is retained in zircaloy.

Wheeler [4.4.10] have measured the rate of diffusion of hydrogen in non-irradiated UO_2 single crystals by using two techniques: (i) vacuum-extraction, and (ii) tritium tracer measurements. The data are fitted with Eq. (4.4.4)

$$D = 0.037 \exp\left[-(14,300 \pm 900) \; cal/RT\right] \quad cm^2 \, s^{-1}. \tag{4.4.4}$$

Since the radius of a hydrogen molecule (0.12 nm) is slightly smaller than that of an oxygen ion (0.14 nm), it is presumed that the hydrogen molecule can enter a vacancy or unoccupied interstitial anion site and diffuse through the structure by the vacancy/interstitial mechanism. This, together with the failure in detection of the ESR signal due to H atoms, suggests that hydrogen is present as molecules in the matrix.

The rate of release of tritium from neutron-irradiated UO_2 either in poly-crystalline or single crystal form has also been measured [4.4.11]. The diffusion coefficients obtained at 500 to 1000 °C are expressed by the equation,

$$D = 0.12(+0.12 - 0.07) \exp\left[(-43,600 \pm 1,800) \; cal/RT\right] \quad cm^2 \, s^{-1} \tag{4.4.5}$$

and are several orders of magnitude lower than those from Wheeler's results. The large discrepancy may be ascribed to different species involved, i.e., either atomic tritium, or molecular hydrogen.

The production of tritium by ternary fission of ^{235}U in a sufficient amount for measurements necessitates a heavy irradiation of the specimen, which will not only put various difficulties in the subsequent experiments, but also allow radiation damage to affect the release behavior of tritium. In order to avoid such difficulties, UO_2 pellets containing 0.1% LiF (UO_2(Li), 64% of the theoretical density) were prepared and neutron-irradiated [4.4.12]. Tritium released during and after the irradiation amounts to less than 1%. The apparent activation energy for the release of tritium is 32 kcal/mol, and heating of the irradiated specimen up to 1050 °C causes almost complete release of tritium as HTO and HT in an approximate ratio 10:1. However, the relative yield of HT increases with the isothermal annealing temperature, and it reaches 32% at 980 °C. From these and additional experimental results, it has been concluded that the formation of HT is primarily ascribed to the partial oxidation of UO_2,

$$UO_2 + x \cdot HTO \rightarrow UO_{2+x} + x \cdot HT. \tag{4.4.6}$$

However, the addition of H_2 or H_2O to the sweep gas leads to a drastic change in the observed tritium distribution; in the former case, more than 90% of the tritium is released as HT, while in the latter, the HTO yield approaches colsely to 100%. These results have been explained by the isotopic exchange and/or isotope dilution.

The characteristics of the tritium release behavior from UO_2 matrix seems to be common for samples with various densities, although the fraction released at any given temperature depends strongly on the density.

Since it is not necessary that the tritium release behavior during the reactor operation, where the specimen is self-heated and always exposed to radiations, should also obey the simple diffusion theory, further in-pile and out-of-pile experiments will be needed for complete understanding of such behavior.

More detailed knowledge on factors such as burn-up, rating, etc. may permit us to predict quantitatively amounts of tritium left during and after the reactor operation.

When irradiated $UO_2(Li)$ was dissolved in binary mixtures of D_2O-DNO_3, both DT and HT were evolved; the DT yield shows DNO_3-concentration dependence and decreases from 6% at 3M to 2% in concentrated DNO_3, while the HT yield remains almost constant ($\sim 2\%$) over the concentration range studied [4.4.13]. The tritium atoms evolved as HT should exist as HT or in its equivalent chemical form in the matrix. It appears that thermalized tritium will combine with oxygen atoms either free or in a combined form, because of the high affinity of oxygen to thermal hydrogen atoms [4.4.14]. Hence, it is likely that HT results from the reaction of energetic tritium atoms in the matrix, although further experiments are needed for confirmation.

Since the release of tritium from cooling water in reactors is dierectly related to its environmental release, determination of the tritium reactions in the water is essential in arranging for its complete removal from the off-gas. Among the nuclear reactions responsible for production of tritium in cooling water, $^6Li(n, \alpha)T$, and $^{10}B(n, n\alpha)T$, and $D(n, \gamma)T$ reactions are the most important. Boron-10 is usually added to the coolant in the form of H_3BO_3 as neutron absorber (chemical shim), and ^6Li in the form of LiOH as pH controller. The T-production through the last nuclear reaction constitutes rather minor fraction in LWR (Light Water Reactor), while it essentially controls the T-concentration of the coolant in HWR (Heavy Water Reactor).

Table 4.4.1. Energetic- and thermal-reaction yields of products in the reaction of recoil T with $H_2O(D_2O)$ (taken from Ref. 4.4.15)

System	Products	Yields (%)		
		Thermal	Energetic	Total
H_2O	HT	~ 1	~ 11	12
	HTO	20	68	88
D_2O	DT	~ 2	~ 11	13
	DTO	19	68	87

In the reactions with water of recoil tritium generated from these nuclear reactions, both HT and HTO are formed in yields of 12% and 68%, respectively [4.4.15]; the former results from the energetic abstraction of H by T, whereas the latter from both energetic and thermal reactions. The apparent isotope effects between H_2O and D_2O are negligible (Table 4.4.1), owing to the counterbalance of the reactivity integral isotope effect by the moderator isotope effect:

$$\alpha H_2O/\alpha D_2O = (I_{HT} + I_{HTO})/(I_{DT} + I_{DTO}), \qquad (4.4.7)$$

where α_i is the average logarithmic energy loss per collision with molecule i, and I_j the reactivity integral for the product j (see Sect. 3.1). Separate experiments have indicated the reactivity isotope effects as 1.60 ± 0.08 [4.4.16].

In the actual reactor operation, however, the tritium released into the off-gas is considered to be practically in the form of HTO. It has been found that the HT from the energetic recoil tritium reactions is deexcited very efficiently in water. HT thus behaves like H_2, concomitantly formed in the radiolysis of water, and is subjected to the attack by OH radicals [4.4.17]. In the absence of any material working as the radical scavenger for OH radicals, HT is easily converted into HTO:

$$HT + OH \rightarrow HTO + H. \tag{4.4.8}$$

Release of Iodine. Friskney et al. [4.4.18] have measured the release rate for various volatile nuclides in the temperature range 700 to 1550 °C. The data were analyzed in terms of the diffusion of rare gases and their halogen precursors. The effective diffusion constants are summarized in Table 4.4.2. Since the measurements have been carried out with a single crystal of natural abundance, the built-up of appreciable plutonium in the surface layer will enhance release of fission products. Thus on irradiation of the single crystal enriched to 20% ^{235}U, diffusion constants obtained become higher by about an order of magnitude. In any cases, however, the diffusion constant for iodine was in a similar range as those for rare gases.

Efforts have also been devoted to investigation of the chemical forms of radio-iodine released from UO_2 [4.4.19−21]. Castleman et al. [4.4.20] have examined the iodine release from U, UC, and UO_2 under He or O_2 atmosphere at high temperature by the use of a thermochromatograph, and quantitatively assigned the chemical forms of iodine. Upon heating [4.4.22] slightly irradiated UO_2 up to 1150 °C at a constant heating rate in the presence of O_2, iodine was released stepwise at about 400, 550, and 750 °C. Organic iodides, mainly in the form of methyl iodide[1], were observed only in the first step. The oxidative pulverization of UO_2 occurred at about 400 °C. The second and third steps, therefore, are characteristic of the release from U_3O_8, and related to the migration of oxygen and uranium vacancies. The chemical identity of radioiodine released from U_3O_8 was examined in either inert or oxygen atmosphere. In the He atmosphere, the radio-iodine species was neither in any form combined with other fission products, nor in elemental form. It is possibly combined with uranium atoms (species A). It reacts easily with O_2 to yield molecular iodine (activation energy, 6.0 ± 0.5 kcal/mol)

---

[1] The formation of organic iodides may be hardly comprehensible to chemists, without addition of any appreciable amounts of organic compounds to the system concerned. However, their formation can be observed whenever one handles with carrier-free radioiodine. For example, on dissolution of irradiated UO_2 in purified nitric acid diluted with distilled water, a few percent of radioiodine involved is evolved as CH_3I. In this particular case, carbon sources may probably be the organic impurities originally present in nitric acid used [4.4.4]. The carbon sources for formation of organic iodides on heating irradiated UO_2 or U_3O_8 have not been identified yet, but presumably are organic fragments resulting from the pyrolysis of organic compounds present around the system, probably on the surfaces of the glass heating tube of the samples.

Table 4.4.2. Diffusion coefficients (cm^2 s^{-1}) calculated for rare gases and their halogen precursors (taken from Ref. 4.4.18)

$10^4/T$	D_{Br}	D_{Kr}	D_I	D_{Xe}
5.8	1.35−11	3.2−14	1.45−13	2.4−13
6.4	1.7−12	6.9−15	2.3−14	1.15−14
7.0	4.0−13	2.1−15	6.8−15	2.3−15
7.6	1.65−13	6.4−16	2.2−15	1.25−15
8.6	5.0−14	1.45−16	2.9−16	7.9−16
9.6	1.65−14	4.5−17	2.7−17	4.7−16

Notation: $4.5-14 \equiv 4.5 \times 10^{-14}$.

[4.4.23]. The organic iodides accompanying the release of iodine into the atmosphere are formed due to radical reactions. Further investigation of the chemical identity of species A using a fission track method may indicate that each iodine atom released is accompanied by 2—4 uranium atoms [4.4.24], suggesting that the released iodine is not in any known chemical combination with uranium atoms, but rather in a "cluster-like" compound which may be formed in situ by the interaction of the iodine with U-containing species (possibly UO_2 or UO_3) released from the matrix.

However, it must be pointed out that fission products with large affinities towards iodine such as Cs also become accumulated in highly burnt-up UO_2. In such cases, iodine is likely to combine with these nuclides.

Fusion Reactors

Although we have the option of various routes to controlled fusion as the source of energy based on thermonuclear reactions, the most promising nuclear reaction at present is

$$D + T = {}^4He(3.25 \text{ MeV}) + n(14.06 \text{ MeV})$$

because of lower temperatures required for the reaction (roughly 1/10 of that for the D—D reaction). Hence, attention is now directed mainly to the use of the D—T fusion reaction. Although the D—T reaction has a disadvantage that it requires fueling with tritium and yields extremely penetrating 14 MeV neutrons, other possible nuclear reactions are also accompanied by side reactions which generate both tritium and neutrons.

Since tritium neither occurs abundantly in nature, nor is manufactured easily in sufficient quantities, D—T fusion reactors will be required most certainly to breed tritium. The use of a lithium-containing substance as blanket material has been the central idea in designing virtually all D—T reactors developed so far, because advantage can be taken of the ${}^6Li(n, \alpha)T$ and ${}^7Li(n, n'\alpha)T$ reactions to breed tritium, provided that reasonable neutron economy is achieved.

Figure 4.4.2 provides a schematic view of the tritium cycle in a D—T fusion reactor. The blanket material chosen must also have the following qualifications: (i) it can moderate fast neutrons, and (ii) it should be a good heat-transfer material.

Although both liquid and solid blanket materials have been widely considered, it has been proposed [4.4.25, 26] that blankets in which lithium is present as solid offer certain advantages over systems containing it as liquid metal, or fused salt. Materials containing lithium, such as LiAl, Li_7Pb_2, $LiAlO_2$, Li_2SiO_3, and Li_2O, are chosen as possible for blanket because they do not yield long-lived induced radioactivities under blanket conditions, the tritium bred is removed from the blanket after a very short average residence time, and their integrities should be kept for a reasonable period of reactor operation.

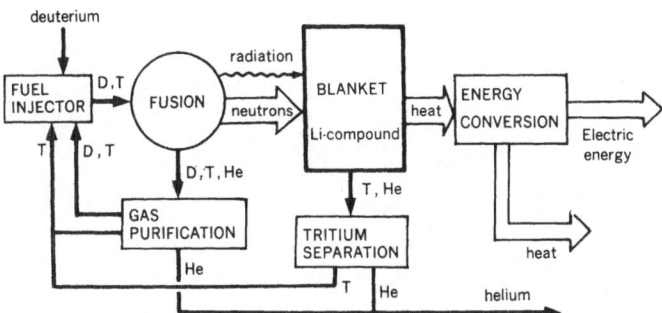

Fig. 4.4.2. Tritium cycle in D—T fusion reactor (taken from Ref. 4.4.37)

As for the chemistry concerned, extensive research will be needed both in the fields related with tritium technology and in the development of blanket constructing materials. Further studies should include not only the recovery, permeation confinement and storage of tritium, but also the radiation damage and compatibility of construction materials.

Tritium Recovery. Wiswall and Wirsing [4.4.38] have tested various lithium compounds for the tritium recovery after neutron irradiation in granular or powder form. The results indicate that low tritium inventory can be attained by continuous extraction of LiAl or Li_7Pb_2 blankets at about 500 °C, and of $LiAlO_2$ or Li_2SiO_3 blankets at about 600 °C.

Lithium oxide has favourable characteristics as blanket material, because of high density and stability towards heat and radiation [4.4.26]. Upon heating neutron-irradiated Li_2O, more than 96% of tritium was released as HTO. Minor fractions were as HT, CH_3T, and homologues with higher molecular weight [4.4.27]. The proposed explanation for the HTO release assumes that tritium atoms formed are first stabilized as LiOT in the Li_2O matrix and then released by the reaction

$$LiOT \cdot LiOH(s) \rightarrow Li_2O(s) + HTO(g).$$

The apparent rate constant for HTO indicates much smaller values both in pre-exponential factor and activation energy than the rate constant for HTO release in the pyrolysis of LiOT [4.4.28]. The discrepancy has been ascribed to the presence of another additional mechanism.

Material Technology. In fusion reactors, the first wall of the blanket is always exposed to a high flux of atoms and ions from plasma. Hence, various species will be released from the wall to the plasma through various processes, such as evaporation due to heat-flux, sputtering, blistering, and desorption of absorbed gases. Since these impurities will not only make the plasma unstable but also lower the plasma temperature, detailed information is needed regarding phenomena which occur at the surface of the first wall, and considerable efforts have been already devoted.

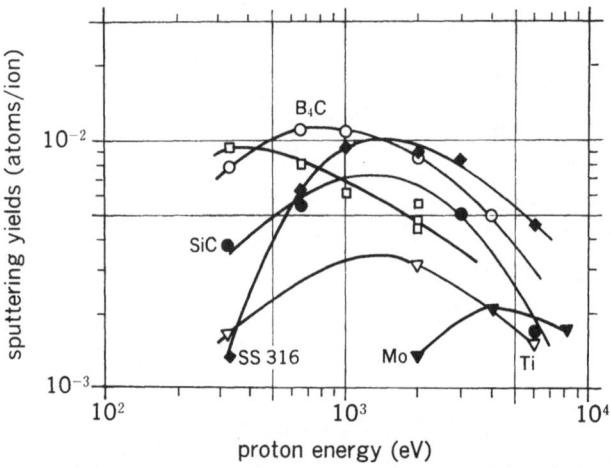

Fig. 4.4.3. Sputtering yields for different materials as a function of proton energy (taken from Ref. 4.4.29). All values are taken at normal incidence and target temperatures between 50 and 150 °C

Bohdansky et al. [4.4.29] have measured the sputtering yield at normal incidence in the bombardment of low atomic number materials with hydrogen ions (Fig. 4.4.3). At higher temperatures, the yield with graphite increased by an order of magnitude due to chemical sputtering. The data for stainless-steel show an increase by a factor of 1.5, probably due to diffusion of different component in the material. The sputtering yield of thin C film deposited on Pt by H_2^+ has also been measured over the range from 20 to 850 °C [4.4.30]. The results indicate that the dose rate increases, the maximum value of the yield decreases, and the temperature at which this maximum occurs increases. Such behaviors are explained in terms of the model of CH_4 production developed by Erents et al. Considering these results regarding the use of carbon in fusion reactors, carbon in any form becomes relatively more attractive as the anticipated wall flux increases.

Recently Wu et al. [4.4.31] have studied gaseous species formed by the reaction of hydrogen on the surface of stainless steel membrane after permeation, as a function of temperature and permeating hydrogen flux. The surface of the stainless steel was coated with or without pyrocarbon. Without the coating, CH_4, CH_3, C_3H_4, and C_4H_6 were released in the temperature range 900 to 1470 °C.

Mostly saturated hydrocarbons with higher molecular weights up to C_6H_{14} were desorbed when the surface was coated with carbon.

The chemical identity of recoil tritium released from the graphite has also become the subject for studies. When recoil tritium arising from the $^3He(n, p)T$ has been allowed to react with graphite in various grades [4.4.32, 33], the tritium atom becomes strongly bound to graphite by forming C−T bond and is not released even at 1000 °C if the graphite was thoroughly degassed and annealed prior to tritium bombardment. Subsequent heating of the samples in the presence of H_2, H_2O, and NH_4 at elevated temperatures leads to the release of up to about 47% of tritium as HT and CH_3T: the HT/CH_3T ratio depends on the temperature and the increase in temperature favors the formation of HT at the expense of CH_3T.

When the irradiated graphite is heated in the presence of acetone vapor above the decomposition temperature of acetone, a burst release of tritium as HT and CH_3T occurs. This is primarily due to the abstraction of tritium from C−T bonds by the radicals produced by pyrolysis of acetone [4.4.33]. The examination of the sample species obtained from the graphite-cooled nuclear reactor seems to indicate that the tritium release temperature becomes higher as the total neutron fluence increases [4.4.34].

The behaviors of hydrogen and helium in solid are relevant to the integrity of the constructing materials. Davis et al. [4.4.35] have determined the depth distribution of tritium and deuterium implanted into titanium and copper. For the determination of tritium, the sample was bombarded with a pulsed proton beam and the energy of neutrons produced by the T(p, n) reaction was measured by the time-of-flight technique. Deuterium depth profiling has been demonstrated by using the D(d, n) reaction. Picraux et al. [4.4.36] have determined the lattice location of deuterium and helium in solid by combining the ion-induced nuclear reactions with ion channeling in single crystals. The results on channeling angular distribution measurements indicate that the implanted deuterium is located in a tetrahedral interstitial site in the bcc tungsten lattice, while the implanted helium is predominantly trapped at a multiple helium plus vacancy sites.

References

[4.4.1]   Lawrence, G. T.: J. Nucl. Mat. *71* 195 (1978)
[4.4.2]   Booth, A. H., Rymer, G. T.: CRDC-720 (1958)
[4.4.3]   MacEwan, J. R., Stevens, W. H.: J. Nucl. Mat. *11* 77 (1964)
[4.4.4]   Numakura, K., et al.: J. Nucl. Sci. Technol. *10*, 367 (1973)
[4.4.5]   Carroll, R. M., Sisman, O. A.: Nucl. Sci. Eng. *21*, 147 (1965)
[4.4.6]   Carroll, R. M., Morgan, J. G., Perez, R. B., Sisman, O.: ibid. *38*, 143 (1969)
[4.4.7]   Miekeley, W., Felix, F. W.: J. Nucl. Mat. *42*, 297 (1972)
[4.4.8]   Matzke, Hj.: Nucl. Appl. *2*, 131 (1966)
[4.4.9]   Matzke, Hj.: Can. J. Phys. *46*, 621 (1968)
[4.4.10] Wheeler, V. J.: J. Nucl. Mat. *40*, 189 (1971)
[4.4.11] Scargill, D.: ibid. *74*, 62 (1978)
[4.4.12] Aratono, Y., Tachikawa, E.: J. Inorg. Nucl. Chem. *39*, 2125 (1977)
[4.4.13] Aratono, Y., Tachikawa, E.: ibid. in press (1981)

[4.4.14] Van Urk, P.: Ph.D Thesis, Univ. of Amsterdam, 1970
[4.4.15] Tachikawa, E., Aratono, Y.: J. Inorg. Nucl. Chem. *38*, 193 (1976)
[4.4.16] Tachikawa, E., Aratono, Y.: Bull. Chem. Soc. Jpn. *48*, 2182 (1975)
[4.4.17] Aratono, Y., Nakashima, M., Tachikawa, E.: J. Inorg. Nucl. Chem. *39*, 1473 (1977)
[4.4.18] Friskney, C. A., Turnbull, J. A., Johnson, F. A., Walter, A. J., Findlay, J. R.: J. Nucl. Mat. *68*, 186 (1977), and unpublished results
[4.4.19] Parker, G. W., Barton, C. J.: *The Technology of Nuclear Research Safety* T. J. Thompson, J. G. Beckerly (eds.) Vol. 2, p. 525, MIT, Massachusetts (1973)
[4.4.20] Castleman, A. W., Jr., Tang, I. N., Munkewitz, H. R.: BNL-9173 (1965)
[4.4.21] Fukuda, K., Handa, M., Shiba, K.: JAERI-M-5846 (1974)
[4.4.22] Tachikawa, E., Saeki, M., Nakashima, M.: J. Inorg. Nucl. Chem. *39*, 749 (1977)
[4.4.23] Tachikawa, E., Nakashima, M.: Int. J. Appl. Rad. Isotopes *28*, 417 (1977)
[4.4.24] Nakashima, M., Tachikawa, E.: J. Inorg. Nucl. Chem. in press (1981)
[4.4.25] Powell, J. R., Miles, F. T., Aronson, A., Winsche, W. E.: BNL-18236 (1973)
[4.4.26] Hiraoka, T.: JAERI-M-5147 (1973).
[4.4.27] Kudo, H., Tanaka, K., Amano, H.: J. Inorg. Nucl. Chem. *40*, 363 (1978)
[4.4.28] Kudo, H.: J. Nucl. Mat. *87*, 185 (1979)
[4.4.29] Bohdansky, J., Roth, J., Sinha, M. K.: Proc. 9th Symp. Fusion Technol., Garmisch-Partenkirchen (1976) p. 541
[4.4.30] Smith, J. S., Meyer, C. H., Jr.: J. Nucl. Mat. *76–77*, 193 (1978)
[4.4.31] Wu, C. H., Kudo, H.: Fusion Tech., 835, (1979)
[4.4.32] Boothe, T. E., Ache, H. J.: J. Phys. Chem. *82*, 1362 (1978)
[4.4.33] Saeki, M., Rowland, F. S.: Abstract, 9th, Int. Symp. Hot Atom Chem. Blacksburg, Virginia, 1977, p. 45
[4.4.34] Saeki, M.: J. Nucl. Mat. in press (1981)
[4.4.35] Davis, J. C., Anderson, J. D., Lefevre, H. W.: UCRL-77153 (1975)
[4.4.36] Picraux, S. T., Vook, F. L.: J. Nucl. Mat. *53*, 246 (1974)
[4.4.37] Kudo, H., Tanaka, K.: Nihon Genshriyoku Gakkaishi *20*, 871 (1978)
[4.4.38] Wiswall, R. H., Wirsing, E. W.: CONF-750989 (1976)

4.5 Current Topics Related to Hot-Atom Chemistry and Future Scope

Let us now look at some current subjects which have not been mentioned in the preceding chapters because they do not appear to be related directly to hot-atom chemistry, but should deserve ample attention of hot atom chemists in seeking for future scope for developments in this field.

As has been pointed out in Sect. 4.1, hot-atom chemistry is essentially related to the production and separation of radioactive nuclides. In fact, hot-atom processes can be used for single-step separation of particular radioisotopes if appropriate target systems and nuclear reactions have been selected. In this connection, there are two other techniques which can also provide means of selective excitation leading to effective isotope separation, i.e., laser and NEET (nuclear excitation by electron transition). Furthermore, hot atom chemists may also be interested in the multi-photon excitation or plasma formation process attained with high-power lasers.

Another subject of interest for hot atom chemists is mesic chemistry. The epithermal reactions of 'hot' muonium atoms with media can be examined by means of μ^+SR technique, or observing spin rotation signal of a muon as a probe. The capture of a negative pion in a condensed phase is usually accompanied by the production of heavy fragments ('star' formation) or short-lived nuclides such as ^{11}C, ^{13}N and ^{15}O, both of which may contribute to medical and biological sciences.

Lasers — Isotope Separation and Hot-Atom Chemistry

In recent years, an increasing attention has been paid to the application of the laser technique to various fields of chemistry. Such applications generally include basic studies of selective or very quick (in 10^{-9} to 10^{-10} s) chemical reactions induced by laser irradiations as well as practical uses in isotope separation and as powerful light sources.

The laser isotope separation consists in irradiating a mixture of isotopes with a monochromatic laser light tuned to excite selectively a particular isotope in the first step, and isolating the excited isotope alone through an irreversible physical or chemical procedure in the second step. For example, laser separation of uranium isotopes ^{235}U and ^{238}U can be performed by two-step photoionization of metalic uranium vapor. At Lawrence Livermore Laboratory, 4 mg of 3% enriched ^{235}U was actually obtained by this process [4.5.1]. An atomic uranium beam prepared by heating uranium metal at $2\,100\,°C$ was irradiated with 378.1 nm Xe ion laser light (1st step) and 350.7/356.4 nm Kr ion laser light (2nd step). The ^{235}U atoms excited selectively by the Xe light were then ionized by the Kr light and separated from ^{238}U in an electrostatic field (Fig. 4.5.1).

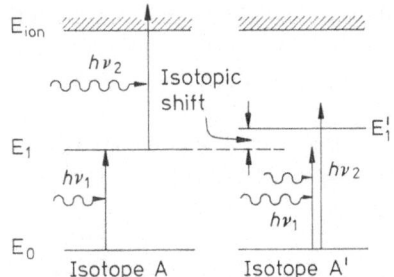

Fig. 4.5.1. Principle of the isotope separation by two-step photoionization

Another possibility for laser isotope separation of uranium is based on the photochemical process with UF_6 [4.5.2]. Since the peaks for $^{235}UF_6$ and $^{238}UF_6$ in the vicinity of 625 cm^{-1} (ν_3) in the infrared spectrum of UF_6 are well resolved below 95 K, the ν_3 mode for $^{235}UF_6$ in supercooled state can be excited selectively by a 16 μ (in wavelength) laser. The excited $^{235}UF_6$ molecules may then be allowed to react with HCl to yield $^{235}UF_5Cl$, which can be isolated as solid on the wall surface. Attempts have also been made to use UF_6 molecules isolated in rare gas matrices at very low temperatures.

A polyatomic molecule (e.g., SF_6) with the vibrational quasi-continuum can be excited to dissociate by absorbing a large amount of energy (a number of infrared photons) in a moderately high infrared field ($\sim 10^6$ W cm^{-2}). Such a multi-photon process is also used for the separation of isotopes because of the selectivity in the lower vibrational transitions due to isotopes. Successful multi-photon separations have been reported for $^{10}B/^{11}B$ in BCl_3 [4.5.3—5], $^{12}C/^{13}C$ in CCl_4 [4.5.6, 7], $^{28}Si/^{29}Si/^{30}Si$ in SiF_4 [5], $^{32}S/^{34}S/^{36}S$ in SF_6 [4.5.8—15], etc.

The multi-photon process should attract attention of hot atom chemists since it provides another tool for generating excited molecules. The energy absorbed by this process is not apparently localized in that particular bond, rather tends to be distributed randomly among the other vibrational modes. Accordingly, the detailed study of the multi-photon process will afford clues to the understanding of the mechanisms for intramolecular transfer of the excitation energy [4.5.16].

There is another possible application of lasers in hot-atom chemistry. It has been found that heating of metal targets with high power (10^{10} to 10^{14} W cm^{-2}) lasers (ruby, Nd-glass, CO_2 lasers, etc.) effectively generates plasma of energetic multiply charged metallic ions [4.5.17—19]. Such metallic ions in the laser plasma are excited species by themselves, and may either be allowed to react with other chemical species, or be used in producing a metal ion beam for implantation studies.

NEET and Isotope Separation

The atomic excitation induced as a consequence of the vacancy creation in the inner shell usually ends up with deexcitation through characteristic X-ray and Auger electron emissions. However, Morita has recently proposed the possibility of another mechanism for atomic deexcitation, i.e., nuclear excitation by electron transition (NEET) [4.5.20]. In this process, the excitation energy is transferred from the atomic electron system to the nucleus which is instantaneously deexcited by γ-transitions. Although the probability for the NEET process is very small as compared with the X-ray and Auger electron emissions, experimental verification of the NEET process has been achieved in some heavy nuclides such as ^{189}Os and ^{237}Np [4.5.21—23]. When a metallic ^{189}Os film was bombarded with 72—100 keV electrons, the conversion electrons emitted from the 6.0-h nuclear state of ^{189}Os (i.e., ^{189m}Os) were detected [4.5.22]. As is shown

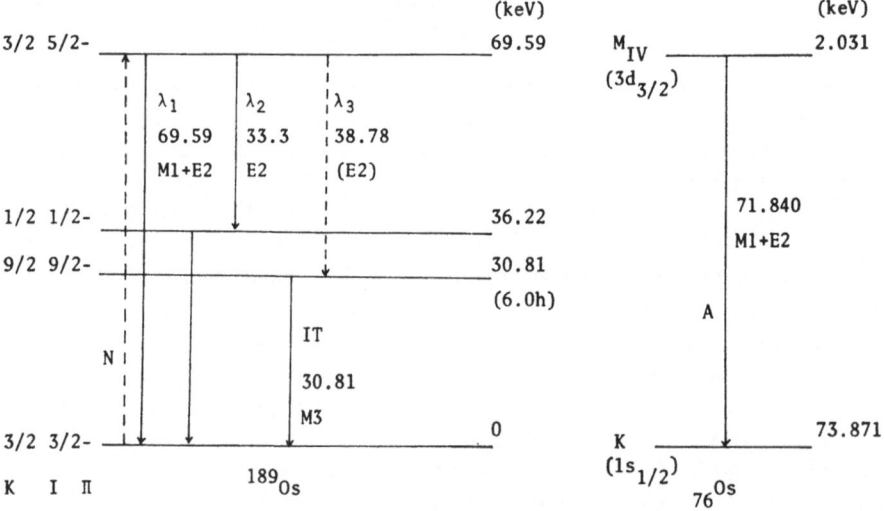

Fig. 4.5.2. Nuclear and atomic levels in ^{189}Os (taken from Ref. 4.5.22)

in Fig. 4.5.2, the creation of a vacancy in the K-shell by electron bombardment excites the ^{189}Os atom, and the atomic deexcitation (A) induces the nuclear excitation to the 69.59 keV state (N) since the transitions A and N have the same multipolarity E2 and nearly equal transition energies. The 30.81 keV state (6.0-h ^{189m}Os) is also populated as the 69.59 keV nuclear state decays in cascades. The threshold energy for the ^{189m}Os production by the NEET is 74 keV, and the cross section at 100 keV is 1.1 nb (Fig. 4.5.3). The contribution of the Coulomb excitation due to the inelastic electron scattering on the ^{189}Os nucleus should be unimportant since the threshold energy for the Coulomb excitation may be below 70 keV and the cross section calculated for that process (broken lines in Fig. 4.5.3) is much smaller than the observed data. The probability for the NEET process in ^{189}Os was estimated as 1.7×10^{-7} based on the measured excitation function.

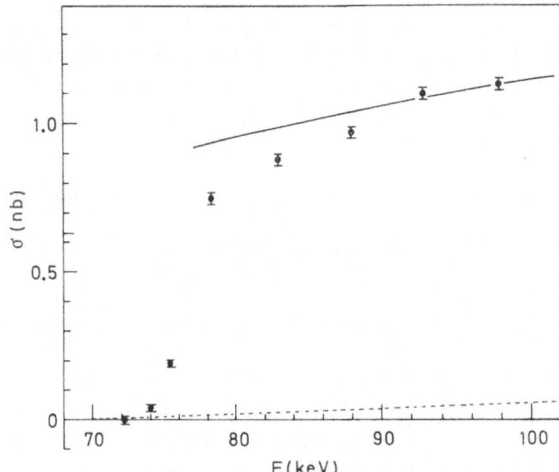

Fig. 4.5.3. Isomer production cross section in ^{189}Os. *Solid circles* represent the observed absolute values, the *solid line* denotes the calculated values for NEET (probability: 1.7 $\times 10^{-7}$), and the *broken line* the values calculated for Coulomb excitation (taken from Ref. 4.5.22)

It will be possible in principle (though not practical yet) to apply the NEET process to uranium isotope separation [4.5.24]. The ^{235}U nucleus has the 30 eV first excited state (26-m ^{235m}U) and 13.1 keV second excited state whereas the first excited state for ^{238}U lies 45 keV above the ground state. If the 26-m ^{235m}U is obtained by exciting the 13.1 keV state alone via the NEET process, hot atom effects associated with the IT decay of ^{235m}U may be utilized for isolating ^{235}U from ^{238}U. It has been suggested that irradiations with lasers or charged particles may greatly improve the yield of ^{235m}U by the NEET [4.5.25].

Mesons and Hot-Atom Chemistry

The π-mesons can be generated in high-energy nuclear collisions, and then the muons (μ-mesons) are formed in the decay of the pi-mesons:

$$\pi^+ \rightarrow \mu^+ + \nu_\mu, \tag{4.5.1}$$

$$\pi^- \rightarrow \mu^- + \bar{\nu}_\mu. \tag{4.5.2}$$

There are now several institutions with large accelerators in the USA, Canada, Switzerland, USSR and Japan, which can generate intense beams of π-mesons and muons for researches in chemistry and life sciences as well as in physics. The muons have especially attracted attention of chemists because they provide a promising nuclear probe for chemical studies [4.5.26, 27]. The muons produced in the decay of pions are 'polarized' with their spins oriented preferentially along their directions of motion. The muons decay with a lifetime of 2.2 µs:

$$\mu^+ \rightarrow e^+ + \bar{\nu}_\mu + \nu_e, \tag{4.5.3}$$

$$\mu^- \rightarrow e^- + \nu_\mu + \bar{\nu}_e. \tag{4.5.4}$$

In the decay of positive muons (μ^+), energetic positrons are emitted with the following angular distribution:

$$W(\theta) = 1 + PA \cos \theta \tag{4.5.5}$$

where P is the degree of polarization, A the decay asymmetry coefficient ($\sim 1/3$), θ the angle between the μ^+ spin and e^+ emission direction. When polarized positive muons are stopped in matter, they decay according to Eq. (4.5.3), precessing in a transverse magnetic field. Hence, the time spectrum for the muon decay observed in a given direction is expressed by

$$N(\theta, t) = N_0 \exp(-t/T_\mu) [1 + PA \cos(\theta - 2\pi ft)] \tag{4.5.6}$$

where T_μ is the muon lifetime (2.2 µs), f the Larmor frequency. For the free muons, the Larmor frequency is

$$f_\mu = 13.554 \text{ kHz/G}. \tag{4.5.7}$$

Figure 4.5.4 indicates a typical time spectrum observed in CCl_4, showing μ^+ precession in 100 G external field [4.5.27]. The muons can be detected by their spin precession, and thus used as a probe for the internal field in the medium. This technique is called µSR in analogy to NMR. Positive muons capture electrons in matter to form neutral atoms, muonium (Mu, or μ^+e^-), which may be regarded

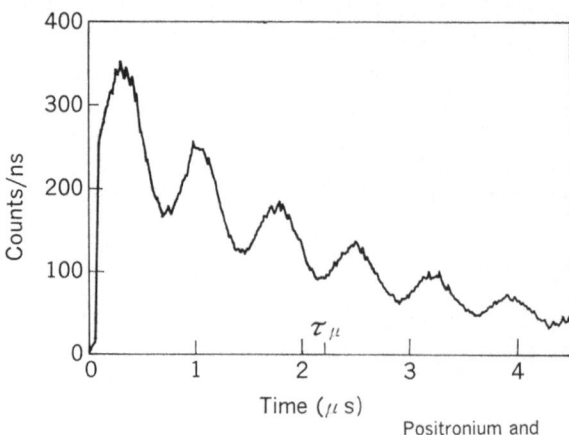

Positronium and
Muonium Chemistry

Fig. 4.5.4. Typical experimental histogram of positive muons stopped in a target of CCl_4 in a magnetic field of 100 gauss. The mean muon decay lifetime $\tau_\mu = 2.20$ µs is seen. (Reprinted with permission from Ref. 4.5.27. Copyright (1979) American Chemical Society)

as a light radioisotope of hydrogen. Since the magnetic moment of Mu is much larger than that of μ^+ (because of the large magnetic moment of electron), the muonium Larmor frequency is

$$f_{Mu} = 1.39 \text{ MHz/G}. \tag{4.5.8}$$

That is, muonium atoms precess much more quickly than muons, even in a weaker magnetic field (this technique is called MSR, corresponding to ESR).

Table 4.5.1. Residual muon polarization data (adapted from Ref. 4.5.27)

Target	P_{res}	Target	P_{res}
CCl_4	1	Acetone	0.54
$SiCl_4$	0.48	C_6H_6	0.15
$SnCl_4$	0.99	C_6H_5Cl	0.23
$TiCl_4$	1.00	C_6H_5Br	0.38
Cyclohexane	0.68	C_6H_5I	0.49
Cyclohexene	0.47	$C_6H_5CH_2Cl$	0.35
1,4-Cyclohexadiene	0.40	$C_6H_5CHCl_2$	0.46
1,3-Cyclohexadiene	0.32	$C_6H_5CCl_3$	0.57
Benzene	0.15	$CHCl_3$	0.85
Hexane	0.62	CH_2Cl_2	0.70
Hexene	0.50	C_6H_5OH	0.38
Hexyne	0.43	Glycerol	0.75
2-Propanol	0.62		

    The chemical behavior of muonium in matter (i.e. muonium chemistry) provides an interesting subject for hot atom chemists and radiation chemists. The energetic polarized positive muons capture electrons in the target medium (liquid phase, as a typical example) to form 'hot' muonium atoms which gradually undergo depolarization. Such hot muonium atoms may either react with the medium (solvent molecules) in the epithermal energy region ($1 \sim 10$ eV) or become thermalized (and possibly react with the medium in the thermal energy region). The epithermal muonium atom reactions take place within a short time range ($< 10^{-9}$ s) and produce diamagnetic compounds in which muon spins sit in diamagnetic environments and never contribute to the MSR signal. However, this fraction of 'hot' muonium atoms in the diamagnetic environments contributes to the μ^+ precession observed at the free muon Larmor frequency. Hence, the contribution of such 'hot' fraction can be expressed in terms of the residual muon polarization, P_{res}, i.e., magnitude of the asymmetry normalized by taking P_{res} for CCl_4 as 1.0 (all muonium atoms stopped in CCl_4 end up in the diamagnetic environment). Table 4.5.1 summarizes typical P_{res} values for various target compounds [4.5.27]. While no obvious correlation is observed between the reactivity towards hot muonium and the strength of the bond being broken, some empirical trends apparently exist between the reactivity and the extent of π-bonding or the number of particular substituting groups in the target molecule [4.5.27]. The hot muonium atoms eventually thermalized without forming diamagnetic compounds correspond to the MSR signal, which undergoes depolarization.

Besides these fractions, muonic radicals (paramagnetic molecules containing the μ^+) have been observed in organic solvents such as benzene and acetone [4.5.28]. They may have been formed in epithermal muonium atom reactions but tend to relax rapidly.

The study of muons and muonium has only started recently but it has already offered a number of poissble chemical applications [4.5.26, 27]. The μSR-MSR resembles the radiotracer technique which is extremely sensitive but will not disturb the systems under study: this is an advantage of the μSR—MSR over conventional means such as NMR and ESR. The μSR will also provide a convenient probe to study the earlier stages of 'hot' muonium reactions as compared with the tritium reactions.

In concluding this section, we may briefly mention another aspect of chemical applications of mesons. Negative mesons (π^- and μ^-) are captured into atomic (or molecular) orbitals to form mesic atoms in the target material [4.5.29]. The mesic X-rays emitted as the negative mesons cascade down to lower atomic orbitals can be used for non-destructive elementary analysis or provide information regarding the chemical states of capturing atoms [4.5.30, 31]. Negative muons (μ^-) captured into the lower atomic orbitals are also regarded as the probe for the magnetic field near the nuclei since their spins also precess in a transverse magnetic field [4.5.32]. As soon as the negative mesons (pions) are eventually captured into atomic nuclei, the highly excited nuclei are broken apart to emit nucleons or heavy fragments (star formation). Thus we can produce short-lived radioisotopes (such as ^{11}C, ^{13}N and ^{15}O) or 'hot' atoms in situ in the target material.

References

[4.5.1] Tuccio, S. A., Foley, R. J., Dubrin, J. W., Krikorian, O.: IEEE J. Quantum Electron., QE-11, 101D (1975)

[4.5.2] Jensen, R. J., Marinuzzi, J. G., Robinson, C. P., Rockwood, S. D.: Laser Focus 12, 51 (1976)

[4.5.3] Ambartzumian, R. V., Gorokhov, Yu, A., Letokhov, V. S., Makarov, G. N., Ryabov, E. A., Chekalin, N. V.: Kvantovaya Elektron. Moscow 2, 2197 (1975)

[4.5.4] Ambartzumian, R. V., Letokhov, V. S.: Sov. Phys. JETP Lett. 20, 273 (1974)

[4.5.5] Lyman, J. L., Rockwood, S. D.: J. Appl. Phys. 47, 595 (1974)

[4.5.6] Ambartzumian, R. V., Gorokhov, Yu. A., Letokhov, V. S., Makarov, G. N., Puretzkii, A. A.: Sov. Phys. JETP Lett. 22, 177 (1975)

[4.5.7] Ambartzumian, R. V., Gorokhov, Y. A., Letokhov, V. S., Makarov, G. N., Puretzkii, A. A.: Phys. Lett., A 56, 183 (1976)

[4.5.8] Ambartzumian, R. V., Gorokhov, Yu. A., Letokhov, V. S., Makarov, G. N.: Sov. Phys. JETP Lett. 21, 171 (1975)

[4.5.9] Ambartzumian, R. V., Gorokhov, Yu. A., Letokhov, V. S., Makarov, G. N.: ibid. 63, 993 (1975)

[4.5.10] Lyman, J. L., Jensen, R. J., Rink, J., Robinson, C. P., Rockwood, S. D.: Appl. Phys. Lett. 27, 87 (1975)

[4.5.11] Aldridge, J. P., Birely, J. H., Cantrell, C. D., Cartwright, D. C.: *Laser Photochemistry, Tunable Lasers and Other Topics, Physics of Quantum Electronics*, Vol. 4, S. F. Jacobs, M. Sargent III, M. O. Scully, C. T. Walker (eds.) Reading: Addison-Wesley (1976), p. 57

[4.5.12] Fuss, W., Cotter, T. P.: Appl. Phys. *12*, 265 (1977)
[4.5.13] Black, J. G., Yablonovitch, Eli, Bloembergen, N.: Phys. Rev. Lett. *38*, 1131 (1977)
[4.5.14] Lin, S. T., Lee, S. M., Ronn, A. M.: Chem. Phys. Lett. *53*, 260 (1978)
[4.5.15] Lyman, J. L., Feldman, B. J., Fischer, R. A.: Opt. Commun. *25*, 391 (1978)
[4.5.16] Grunwald, E., Dever, D. F., Keehn, P. M.: *Megawatt Infrared Laser Chemistry*, New York: Wiley-Interscience (1978)
[4.5.17] Linlor, W. I.: Appl. Phys. Lett. *3*, 210 (1963)
[4.5.18] Langer, P., Tonou, G., Floux, F., Ducause, A.: J. Quant. Electronics, QE-2, 499 (1966)
[4.5.19] Ehler, A. W.: J. Appl. Phys. *46*, 2464 (1975)
[4.5.20] Morita, M.: Prog. Theor. Phys. *49*, 1574 (1973)
[4.5.21] Otozai, K., Arakawa, R., Morita, M.: ibid. *50*, 1771 (1973)
[4.5.22] Otozai, K., Arakawa, R., Saito, T.: Nucl. Phys. *A 297*, 97 (1978)
[4.5.23] Shinohara, A., Saito, T., Arakawa, R., Otozai, K., Baba, H., Hata, K., Suzuki, T.: Submitted to JAERI-M Report
[4.5.24] Otozai, K.: in *Nuclear Phenomena and Analytical Chemistry* (Chemistry Reviews Series 29), T. Tominaga (ed.). Chemical Society of Japan (1980), p. 135
[4.5.25] Okamoto, K.: J. Nucl. Sci. Tech. *14*, 762 (1977)
[4.5.26] Brewer, J. H., Crowe, K. M.: Ann. Rev. Nucl. Sci. *28*, 239 (1978)
[4.5.27] Fleming, D. G., Garner, D. M., Vaz, L. C., Walker, D. C.: in *Positronium and Muonium Chemistry* (Advances in Chemistry Series 175), H. J. Ache (ed.), American Chemical Society (1979)
[4.5.28] Roduner, E., Percival, P. W., Fleming, D. G., Hochmann, J., Fischer, H.: Chem. Phys. Lett. *57*, 37 (1978)
[4.5.29] Ponomarev, L. I.: Ann. Rev. Nucl. Sci. *23*, 395 (1973)
[4.5.30] Knight, J. D., Orth, C. J., Schillaci, M. E., Naumann, R. A., Daniel, H., Springer, K., Knowles, H. B.: Phys. Rev. *A 13*, 43 (1976)
[4.5.31] Mausnar, L. F., Naumann, R. A., Monard, J. A., Kaplan, S.: Phys. Rev. *A 15*, 479 (1977)
[4.5.32] Nagamiya, S., Nagamine, K., Hashimoto, O., Yamasaki, T.: Phys. Rev. Lett. *35*, 308 (1975)

Subject Index

Anwendung von Isotopen in der Organischen Chemie und Biochemie

1. Band: H. Simon, H. G. Floss
Bestimmung der Isotopenverteilung in markierten Verbindungen
1967. 5 Abbildungen. X, 247 Seiten
ISBN 3-540-03726-8

Inhaltsübersicht:
Allgemeines: Einführung. – Carbonsäuren. – Aliphatische Kohlenwasserstoffe. – Alkohole, Amine und Halogenverbindungen. – Kohlenhydrate. – Abbau aromatischer Ringe. – Cycloaliphatische Verbindungen, Isoprenoide, Steroide. – Abbau heterocyclischer Ringe. – Spezielle Naturstoffe. – Verbindungsliste. – Literatur. – Sachregister.

2. Band:
Messung von radioaktiven und stabilen Isotopen
Von P. Rauschenbach, H.-L. Schmidt, H. Simon, R. Tykva, M. Wenzel
Herausgeber: H. Simon
1974. 87 Abbildungen. XIII, 430 Seiten
ISBN 3-540-06587-3

Inhaltsübersicht:
Allgemeines und Prinzipien der Radioaktivitätsmessung. – Parameter, die auf Genauigkeit und Reproduzierbarkeit von Einfluß sind. Fehlerbetrachtung. – Präparation der Proben und deren Messung. – Die Bestimmung geringer Radioaktivität – Messung mehrfachmarkierter Proben. – Radiochromatographie. – Analyse von stabilisotop markieten Verbindungen.

A. F. Williams
A Theoretical Approach to Inorganic Chemistry

1979. 144 figures, 17 tables. XII, 316 pages
ISBN 3-540-09073-8

Contents: Quantum Mechanics and Atomic Theory. – Simple Molecular Orbital Theory. – Structural Applications of Molecular Orbital Theory. – Electronic Spectra and Magnetic Properties of Inorganic Compounds. – Alternative Methods and Concepts. – Mechanism and Reactivity. – Descriptive Chemistry. – Physical and Spectroscopic Methods. – Appendices. – Subject Index.

This book is intended to outline the application of simple quantum mechaniscs to the study of inorganic chemistry, and to show its potential for systematizing and understanding the structure, physical properties, and reactivities of inorganic compounds. The considerable development of inorganic chemistry in recent years necessitates the establishment of a theoretical framework if the student is to acquire sound knowledge of the subject. An effort has been made to cover a wide range of subjects, and to encourage the reader to think of further extensions of the theories discussed. The importance of the critical application of theory is emphasized, and, although the book is concerned chiefly with molecular orbital theory, other approaches are discussed. The book is intended for students in the latter half of their undergraduate studies.

Springer-Verlag Berlin Heidelberg New York

Inorganic Chemistry Concepts

Editors: M. Becke, C. K. Jørgensen, M. F. Lappert,
S. J. Lippard, J. L. Margrave, K. Niedenzu,
R. W. Parry, H. Yamatera

Springer-Verlag
Berlin
Heidelberg
New York